Magnetic Resonance in Food Science
Latest Developments

Magnetic Resonance in Food Science
Latest Developments

Edited by

P.S. Belton
University of East Anglia, Norwich, UK

A.M. Gil
University of Aveiro, Aveiro, Portugal

G.A. Webb
University of Surrey, London, UK

D. Rutledge
Institut National de la Recherche Agron, Paris, France

RS•C

advancing the chemical sciences

The proceedings of Magnetic Resonance in Food Science: Latest Developments held in Paris on 4–6 September 2002.

Special Publication No. 286

ISBN 0-85404-886-3

A catalogue record for this book is available from the British Library

Published by The Royal Society of Chemistry,
Thomas Graham House, Science Park, Milton Road,
Cambridge CB4 0WF, UK
Registered Charity No. 207890

For further information see our web site at www.rsc.org

Printed by Athenaeum Press Ltd, Gateshead, Tyne and Wear, UK

Preface

The Institut National Agronomique Paris-Grignon (INA-PG) was the venue of the sixth in the series of International Conferences on Applications of Magnetic Resonance in Food Science. It was held between the 4th and the 6th of September 2002 and attracted 130 registered participants from 28 countries with representatives from all five of the continents. Like previous members of this series the sixth conference was a very successful one. There are perhaps many reasons for this. The relatively compact size of the conference leads to a ready interchange of ideas and views between participants. In addition the wide range of techniques: spectroscopy, relaxation and imaging, covered by the term Magnetic Resonance continue to evolve in many interesting directions leading to various new applications in food science. Food science itself covers many aspects of chemistry and physics including molecular structure and reactivity, analysis and rheology.

The conference was divided into four symposia; Symposium I covered "Magnetic Resonance – The Developing Scene", Symposium II dealt with "Food: The Human Aspect", "Food: Structure and Dynamics" was covered in Symposium III, while Symposium IV related to "Food Quality Control". Each of the symposia consisted of major and minor oral contributions which is reflected in the various lengths of the chapters contained in this volume. In addition to the oral presentations the conference also attracted 74 posters.

The first eleven chapters relate to Symposium I, the next six chapters derive from Symposium II, the following five chapters cover material presented in Symposium III and the final six chapters contain topics covered by Symposium IV. In addition to the presentations which gave rise to the work recorded in this volume there was a lecture by the recent Nobel laureate, Kurt Wuethrich, on "NMR, Structural Biology of Prion Proteins and the BSE Crisis".

The editors wish to thank the authors for the prompt submission of their camera-ready copy manuscripts. In addition thanks go to the production staff of the Royal Society of Chemistry for much helpful advice during the production of this volume.

Contents

Food: The Human Aspect

Food: Structure and Dynamics

Food Quality Control

Magnetic Resonance – The Developing Scene

THE FOOD SUPPLY CHAIN: THE PRESENT ROLE AND FUTURE POTENTIAL OF NMR

M.J. Gidley, S. Ablett and D.R. Martin

Unilever R + D Colworth, Sharnbrook, Bedford, MK44 1LQ, UK

1 INTRODUCTION

The processes involved in the growing, processing, distribution, and eating of food around the world are both complex and diverse. Factors such as local customs and climates, culinary heritage and traditions, interest in nutrition and health, affordability, and availability of storage and distribution networks are all variable, and affect the relative importance of food-related questions and how they are addressed. A simplified supply chain provides a convenient framework for discussing the wide range of issues related to food. This framework also serves to categorise the diverse range of applications that NMR techniques currently address, as well as the potential for further exploitation of this most versatile and incisive analytical methodology. This overview will follow the various stages of the food supply chain as shown below, exemplifying some of the ways in which NMR techniques can be applied.

Raw Materials - Manufacture - Storage/Distribution - Eating - Digestion/Metabolism

Scheme 1 *A (very) simplified food supply chain*

NMR has an important scientific role to play in each stage of this supply chain, with the whole panoply of magnetic resonance approaches finding application at specific stages. Five types of NMR experiment have found particularly widespread use in studies of foods, and these will form the basis for most of the data reviewed here. *High resolution 'solution state'* methods are particularly suited to defining molecular structures, often in exquisite detail, and increasingly in complex mixtures. Molecular complexity is an inherent characteristic of food with many examples of NMR analysis of multi-component raw materials, and the recent development of metabolite analysis in human fluids providing links from the beginning to the end of the food chain. *High resolution 'solid state'* methods have been applied to many of the bulk structuring systems of food (carbohydrate, oils, proteins), and can also be used to define molecular structures within complex matrices such as biological raw materials and assembled food structures. *Relaxation time* measurements are particularly suitable as a probe of water and oil micro-environments within the multi-compartment structures characteristic of both natural materials and assembled foods. *Diffusion* methods take advantage of magnetic field gradients and are useful in defining both bulk diffusion and molecular self diffusion rates non-invasively. *Imaging* methods are applied increasingly to the non-invasive visualisation of both natural materials and assembled foods, with the information content growing as hardware and image contrast methods improve. The area of in-body imaging of brain and digestive system response to food is a particularly promising and challenging one for the future.

The following sections discuss some of the issues related to each stage of the supply chain (Scheme 1) with examples of NMR approaches that have been found to be of use. Previous volumes in this series[1,2] provide many further examples.

2 RAW MATERIALS

2.1 Molecular composition

The raw materials for food are the products of agriculture, either directly or indirectly. The inherent molecular and structural complexity / heterogeneity of natural materials pose many challenges for analytical technologies. In principle, NMR analysis techniques have a big advantage in that they are non-prejudicial. That is target molecules or structures do not need to be pre-selected for analysis, as all relevant nuclei are affected by the experiment and can in principle be detected. In practice, NMR is a frustratingly insensitive technique compared for instance with mass spectrometry. The reason for this can be understood using an energy levels description. Inherently, nuclear spin states have only a small difference in energy between ground and excited states. This leads to nearly equal population of spin states in the absence of a magnetic field, and consequently only a minor difference in populations under resonance conditions. The difference in population between ground and excited states essentially drives all NMR methods. The fact that only a very small minority of individual spins are perturbed by the experiment is one of the main reasons for the inherent insensitivity of NMR, compared to many other spectroscopic methods where energy gaps are typically much greater.

Despite the sensitivity limitations, the ability to monitor (non-invasively) all abundant molecules in a raw material is a major driver for NMR applications. This does not need to involve identification of every resonance, as pattern recognition and related techniques for discriminating between data sets have developed rapidly over the last few years. Such approaches are very useful for determining authenticity when no genetic material is present (otherwise genetic fingerprinting is likely to become the technique of choice). There are many examples of NMR applications in authenticity, some involving elegant isotope quantification methods to establish biological origin. The ultimate expression of molecular analysis methods is in the relatively new discipline of metabolomics, which has as its target the description of all molecular components in a system. As genomic and proteomic analyses of crop plants and farm animals become more common, there is likely to be a surge of interest in metabolomic technologies. NMR methods will have a major role to play here.

These approaches assume that all molecules of interest can be observed. This is not always the case for classical high resolution NMR, because for example the molecules in question have inappropriate relaxation properties (due e.g. to rigidity or paramagnetism), or because they are present in a microenvironment that prevents isotropic movement (e.g.

in cellular compartments). If molecular rigidity is the key characteristic, then the techniques of 'solid state' NMR can be applied (see below). If mobile molecules are present in a microenvironment that prevents isotropic reorientation, this can (in principle) be solved using Magic Angle Spinning as this geometrically averages such anisotropies to zero. This approach has found many applications recently with the availability of commercial ^{1}H HR-MAS probes. An example is shown in Fig 1 : this compares the ^{1}H NMR spectra for tea leaves obtained under 'conventional' and MAS conditions, clearly illustrating the advantage of the MAS technique for observing low molecular weight solutes within biological material.

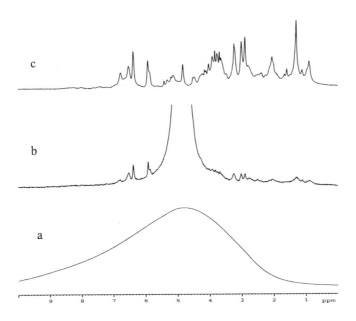

Figure 1 *a : conventional 'high resolution' ^{1}H NMR spectrum of tea leaf*
b : magic angle spinning ^{1}H NMR spectrum of tea leaf
c : as (b) with water suppression

The resolution achievable with commercial MAS probes is sufficient to make many assignments to molecular components of tea leaves (Fig 2). This signal resolution is starting to approach that which can be obtained from true solutions (Fig 3), and represents a major step forward in technology over the last few years. Applications of MAS 'solution state' NMR spectroscopy can be foreseen for a wide range of (non-invasive) molecular analysis needs within complex food or biological matrices.

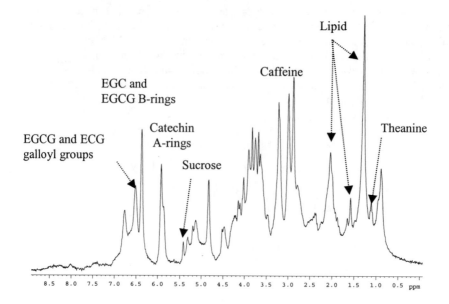

Figure 2   *¹H HRMAS spectrum of a fresh tea leaf.*

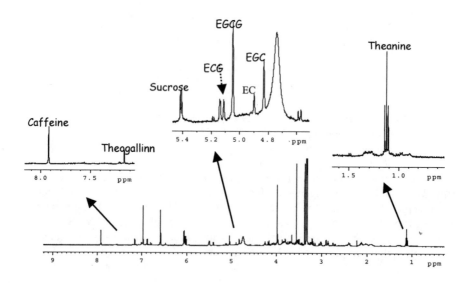

Figure 3   *¹H High resolution spectrum of a green tea infusion.*

2.2 Crystalline and Molecular Order in Solids

Many of the characteristic components of food are sufficiently 'solid-like' that they possess rapid NMR relaxation properties that prevent the use of 'solution state' NMR methods. However, the approaches of 'solid state' NMR are applicable[3]. In particular, the use of cross polarisation excitation (from ^{1}H to ^{13}C), high power ^{1}H decoupling (to inhibit relaxation pathways), and magic angle spinning (to reduce chemical shift anisotropy effects) has found widespread use. This experiment (normally referred to as ^{13}C CPMAS) has been applied to all common food components with solid character. Examples include starch[4], oils[5,6], cellulose[7], and plant cell walls[8]. In addition to characterising primary structures, this experiment has the unique advantage of providing a fingerprint of organised molecular structure. For example, crystalline polymorphs of starch and triglycerides characterised by X-ray diffraction give recognisably different ^{13}C CPMAS spectra[4,6]. For cellulose, NMR has been found to be the method of choice for characterising crystal polymorphs, as spectral differences between I alpha and I beta forms are more obvious from NMR spectra than from diffraction data[7].

NMR and diffraction methods give complementary structural information as they report on different distance scales. For diffraction to be detected, there needs to be a conserved repetitive structure over several tens of nm (depending on the size of the unit cell). By contrast, NMR chemical shifts are sensitive to short-range (nm) effects. This means that all 'crystalline' order detected by (X-ray) diffraction should correspond to 'crystalline/ ordered' signals in the CPMAS spectrum. In addition, however, NMR reports on ordered structures that are either too short range or not arranged perfectly enough to show significant diffraction. As an example, analysis of starch granules shows a level of (X-ray detected) crystalline order of 20-30%, and a level of short range (NMR detected) molecular order of 40-50%[4].

A further example of this distinction between short-range 'molecular order' (as detected by ^{13}C CPMAS NMR) and long-range 'crystalline order' (as detected by X-ray diffraction) is

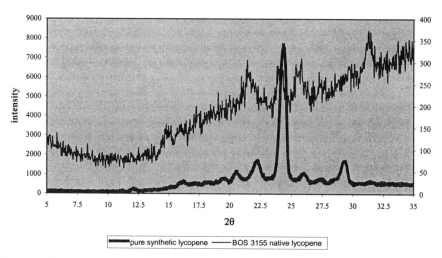

Figure 4 *Powder X-ray diffraction patterns for lycopene isolated from tomatoes (upper trace) or synthesised chemically (lower trace)*

illustrated by a recent study of lycopene[9], the main origin of the red colour in tomatoes and a potent antioxidant with potential health benefits. In tomatoes, lycopene is synthesised in relatively small amounts in specialised organelles (chromoplasts) in what is presumed to be a crystalline form, as 'crystals' are observed microscopically. However, isolation of purified lycopene from tomatoes using mild procedures results in only very weak X-ray diffraction (Fig 4). This is in contrast to lycopene produced by chemical synthesis, which clearly displays a high level of crystallinity (Fig 4). The modest crystallinity of native lycopene could either be due to a non-organised arrangements of molecules, or to the presence of small domains of organisation that are not large enough to diffract X-rays efficiently. [13]C CPMAS spectra (Fig 5) show similar chemical shifts and signal widths for synthetic and natural lycopene. Chemical shifts are distinct from those found in solution[9].

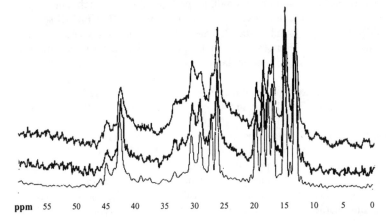

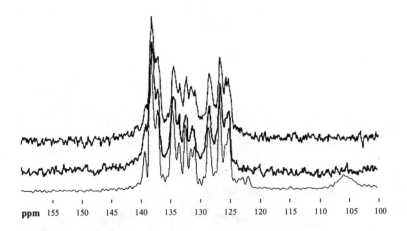

Figure 5 *Two regions of ^{13}C CPMAS NMR spectra for lycopene preparations from tomatoes (upper two traces in each set) and synthetic lycopene (lower trace in each set)*

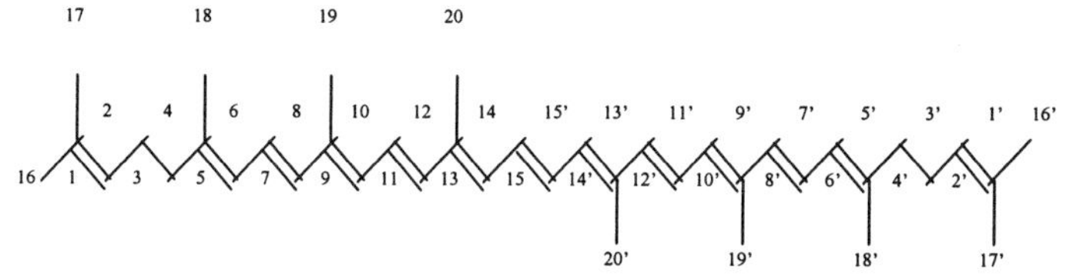

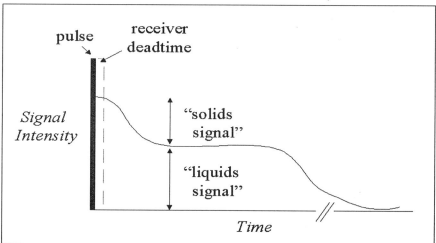

Figure 6 *Molecular structure of lycopene. Signals for carbons 3,4,3',4', and all methyl groups are in the region 10-50 ppm. All other signals are at 120-140 ppm (Figure 5).*

Taken together, these data show that lycopene has a characteristic short-range molecular organisation irrespective of whether it is produced synthetically or by Nature. This presumably reflects a 'stacking' of the symmetrical and highly hydrophobic lycopene molecule. In Nature, this organised molecular assembly is of sufficiently limited size to give only weak diffraction effects. In contrast, synthetic lycopene has the same (or very similar) molecular organisation repeated over a longer distance scale, giving rise to the observed diffraction (Fig 4). It is not known whether this difference between synthetic and natural lycopene has any consequence for bioavailability or potential health benefits.

2.3 Mixtures of solids and liquids

Many foods contain both 'solid' and 'liquid' components. Whilst individual phases can be characterised in mixtures using the techniques discussed above, it is also important to determine the relative amounts of constituent phases. For the specific, but important, case of solid/liquid fat mixtures, NMR relaxation has become the analytical method of choice.

Figure 7 *Schematic representation of (^1H) signal decay in a mixed solid + liquid fat system. The large difference in relaxation timescale between the two components makes quantification straightforward.*

The principle of the approach is illustrated in Figure 7. The total signal intensity (typically measured after $10\mu s$) is proportional to the total number of protons present (in both solid + liquid phases). After an appropriate time delay (typically $70\mu s$), essentially all the fast-relaxing 'solid' fat protons ($T_2 \sim 10\mu s$) have decayed. Residual intensity is then due to protons from 'liquid' fat which have T_2 values of > 1 millisecond. Thus, the solid/liquid ratio can be established[10]from the two intensities. In practice, the effect of spectrometer deadtime has to be taken into account in establishing the total proton intensity. This is done by calibration against standards of known solid/liquid ratios.

3 MANUFACTURE

An increasing proportion of food is manufactured. Sometimes this is to produce food forms that are difficult to prepare in the kitchen e.g. ice cream, margarine. For other foods, manufacturing provides preservation through e.g. heating, drying or freezing. In addition, there is a growing trend for food to be eaten out-of-home: this often involves pre-manufactured components.

3.1 Freezing

Freezing provides a typical example of food processing/manufacture. The benefits of freezing include the preservation of flavour and nutritional benefits over an extended time period, allowing for example, the availability of otherwise seasonal crops all year round. A continual issue with freezing is how to preserve native macroscopic structures, in particular how to avoid ice crystal damage within delicate food structures. NMR relaxation of the water component in foods provides a very useful probe for the effects of freezing on food structure. An example is shown in Figure 8. In native fish muscle, water is contained within relatively small pores in muscle fibres. For some fish muscle, freezing and thawing results in larger pools of water between muscle fibres. As the ^1H T_2 value of water decreases by molecular collision/exchange with the surfaces of pores, an increase in pore size results in an increase in T_2 value, as illustrated below[11].

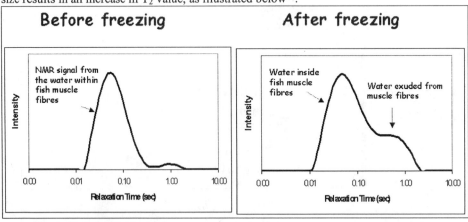

Figure 8 *Freezing/thawing of fish muscle results in an increased proportion of T_2 values >0.1 sec, reflecting water 'pools' formed by ice damage[11].*

Although this post-mortem type of study is useful, it would be preferable to be able to study the formation of ice directly. This is where magnetic resonance imaging can help. Fig 9 shows MR images as a function of time for a cylinder of blanched potato undergoing a freezing process. The non-frozen water and ice have very different relaxation properties, such that the ice is not detected. The images show a loss of signal intensity (i.e. non-frozen water) from the edge of the image inwards, suggesting progressive ice formation from the periphery to the centre of the cylinder. Although the resolution is not great enough to pick up effects at the single cell level (ca 100 microns), MRI provides a powerful addition to the arsenal of tools available to study freezing phenomena.

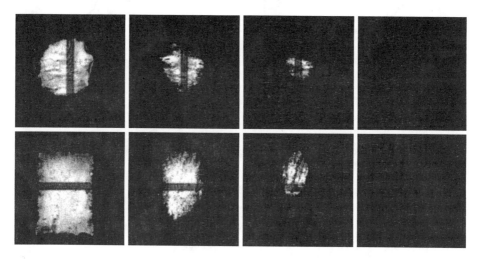

Figure 9 *Orthogonal Gradient Echo images of a cylinder of blanched potato (8mm diameter, 12mm height). Upper set - top view. Lower set - side view. The left hand pair of images show the unfrozen cylinder: freezing time increases for sets of images from left to right.*

3.2 Emulsification

Many processed foods contain both aqueous and lipid components, and frequently these are present as an emulsion. Droplet size within emulsions is a major quality determinant for a range of different foods. NMR self diffusion measurements provide a key non-invasive technique for the sizing of water droplets within emulsions. The principles of the technique are outlined in Figs. 10 and 11. Two equal field gradient pulses are applied within a standard spin-echo pulse sequence, where the intensity of the observed echo is reduced by the effects of molecular diffusion. In the case of free diffusion, the echo intensity is diminished as the effects of diffusion are increased, such as longer diffusion times (Δ) or increasing gradient strength. In the case of droplets however molecular diffusion is 'restricted' by the boundary of the droplet, and once this boundary is reached the echo intensity is no longer reduced with increasing diffusion times. By monitoring signal attenuation as a function of gradient strength, droplet size can be determined[12]. This method is suitable for characterising water droplet sizes in the order of 1-50 microns i.e. the typical size range found in food emulsions.

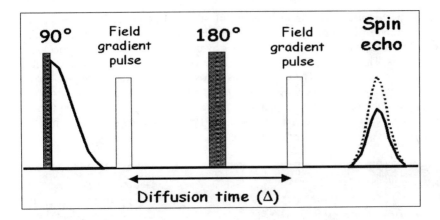

Figure 10 *Pulse sequence for NMR diffusion experiment.*

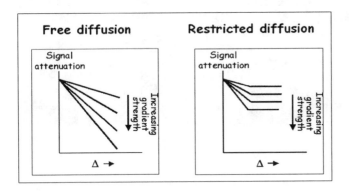

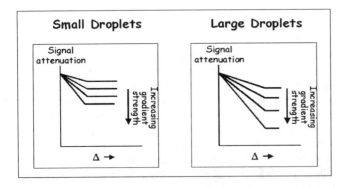

Figure 11 *How varying gradient strength can be used to probe diffusion (upper panel) and measure droplet size (lower panel).*

4 STORAGE & DISTRIBUTION

Although food is often highest quality if eaten immediately after harvesting or preparation, in reality a majority of food undergoes some storage or distribution, during which changes can occur. Sometimes these are beneficial (e.g. ageing of cheese, wine etc), but many times changes are deleterious. This is often a reflection of the fact that most food is in a metastable or kinetically-trapped state. Given sufficient time and appropriate conditions, equilibration is inevitable. One obvious example is the penetration of water e.g. from the surface into the bulk of a solid food. The non-invasive and location-resolving features of MRI can be used to track and quantify such events, and thereby provide an objective measure of e.g. the effectiveness of potential barrier materials. Figure 12 illustrates both the principle and a specific example of a 1D version of imaging that can be used on standard benchtop spectrometers equipped with pulsed field gradient coils.

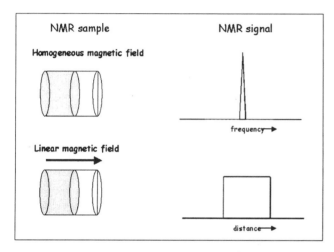

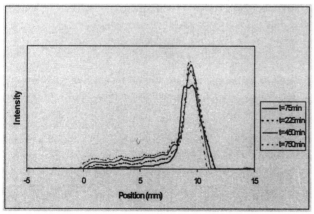

Figure 12 *Upper panel : principle of 1D MRI profiling.*
Lower panel : time course of water penetration from a spread('8-11' mm at start) into a dry cracker of 8 mm depth. Penetration throughout the cracker is seen after a few hours.

Quantitative information on the location of water as a function of time therefore provides a convenient non-invasive measure of the rate of bulk diffusion of water within this system.

Crystallisation is another example of an equilibration event that can occur on storage. Examples include triglycerides, starch ('retrogradation'), and as illustrated below, sucrose. Sucrose can be obtained in a 'glassy' state by freeze drying an aqueous solution. Heating of this material above the effective glass transition temperature results in crystallisation. This process can be monitored using ^{13}C CPMAS NMR as illustrated in Fig 13. At low temperatures, broad spectral lines are indicative of the wide range of local conformations that are present in the kinetically 'frozen' glassy state. The 'average' chemical shift for individual signals corresponds to the sharp signals seen in aqueous solution. Above the effective glass transition (ca 40°C), a series of sharp lines with different chemical shifts start to be observed. At 60°C, this form predominates, and is identical to the spectrum of crystallised sucrose.

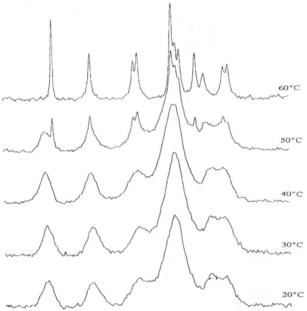

Figure 13 *Effect of increasing temperature on 'glassy' sucrose as monitored by ^{13}C CPMAS NMR. The effective glass transition temperature (as determined by DSC) is ca 40°C.*

5 EATING, DIGESTION & METABOLISM

NMR has found little application (so far) to the direct study of the eating process. This is in part due to the non-availability of vertical bore whole body imaging magnets coupled with the dynamic nature of the process. Imaging is however playing an increasingly important role in monitoring the response of both the brain and the stomach to food. 'Functional' MRI monitors local changes in blood flow and oxygenation in the brain through effects on T_2 values. This leads to the identification of zones within the brain that 'light up' following a stimulus such as eating[13]. As knowledge of brain function increases, it may be possible to use fMRI to reveal relationships between food type/quality/perception and feelings of sensorial pleasure, satiety, mood etc. in an objective manner.

Imaging techniques are also shedding new light on the processes that occur in the stomach during digestion[14]. The mixing and disruption of solid foods can be monitored directly, allowing hypotheses on the rheological conditions in the stomach to be derived. The relatively slow mixing of 'bulky' foods such as porridge, and the relative intactness of vegetable tissues in the stomach exemplify the type of insights that can be obtained. The stomach is a relatively large organ with limited movement within the body. In contrast, the small intestine is relatively narrow and moves significantly. These factors make it currently very difficult to use MR imaging for stages of digestion beyond the stomach.

Many of the constituents of food are taken up by the body, and transported to various organs. As mentioned earlier in the context of raw material characterisation, NMR has a leading role to play in the definition of molecular composition in complex mixtures. Analysis of biofluids such as blood, cerebrospinal fluid, and urine provide good examples of the power of this approach[15,16], and are discussed elsewhere in this volume. Such analyses provide fingerprints of the response to e.g. food, and potentially provide a way of characterising diet-gene interactions (i.e. understanding why different people have different responses to food).

6 FUTURE PROSPECTS

NMR will remain the single most versatile and informative spectroscopic method for food chain investigations. The complementary nature of (a) the physical information that can be obtained from relaxation-related methods, (b) the chemical structure (primary, secondary, and tertiary) insights from high resolution methods, and (c) the spatial information from imaging, provides a wonderful toolbox. Drivers for future improvements in capability are:

- the ingenuity of experimenters
- the ever-increasing sophistication of software and data analysis tools
- improvements in NMR hardware
- exploitation of hyphenated techniques (e.g. LC-NMR, LC-NMR-MS)
- increased robustness and reduced cost of MRI and localised spectroscopy
- the pressure from post-genomic biology for detailed metabolomic analyses

At the leading edge, there will undoubtedly be a continuing expansion of new NMR applications. However, the food industry is not as well placed financially as e.g. the medical/pharmaceutical sector to support expensive equipment. A real challenge for the further exploitation of NMR within the food supply chain is therefore to develop robust, portable and economical equipment that can be used not only in fields and factories, but also with consumers. The risk of not doing so is that NMR becomes an elitist tool that does not deliver its potential to address the needs of the entire food chain.

There will always be limitations of NMR, most notably in sensitivity. NMR can never approach the sensitivity of mass spectrometry for molecular analysis. Furthermore, when high resolution (i.e. chemical shifts) and/or spatial information are required, field homogeneity becomes critical. This inevitably leads to both bulky and expensive equipment. A further specific example of a current limitation is the geometry of standard MRI instruments. When used to image people, this requires the subject to lie down. This is not the normal position for people either during or soon after eating!

7 CONCLUSIONS

The food supply chain is central to both the health and wealth of individuals and nations. From a consumer perspective, there are increasing demands for guaranteed safety, authenticity/naturalness of raw materials and ingredients, minimal processing, instant availability, exciting/satisfying tastes and textures, and the promise of a prolonged and healthy life. In addition to addressing all these factors, the agri-food industry also focuses on achieving optimal benefit/cost ratios, stability during storage and distribution, and responding to regulatory pressures on labelling, contaminants, and (increasingly in the future) sustainability. This amazingly diverse set of challenges requires the best analytical tools to support innovation, identify technical opportunities, and monitor products throughout the food chain. NMR will remain the single most important analytical technique in food chain applications, through the diversity of experiment and the breadth and depth of information that can be obtained.

Acknowledgements

The authors would like to thank Adrienne Davis, Arthur Darke, Rebecca Lipscombe and Luisa Gambelli for help with acquiring data or providing samples.

References

1   *Magnetic resonance in food science: a view to the future*, eds. G.A. Webb, P.S. Belton, A.M. Gil and I. Delgadillo, Royal Society of Chemistry, Cambridge, 2001.
2   *Advances in magnetic resonance in food science*, eds. P.S. Belton, B.P. Hills and G.A. Webb, Royal Society of Chemistry, Cambridge, 1999.
3   M.J. Gidley, A.J. McArthur, A.H. Darke and S. Ablett, 'High resolution NMR spectroscopy of solid and semi-solid food components' in *New physico-chemical techniques for the characterisation of complex food systems*, ed. E. Dickinson, Blackie Academic and Professional, Glasgow, 1995, pp 296-318.
4   M.J. Gidley and S.M. Bociek, *J. Am. Chem. Soc.*, 1985, **107**, 7040.
5   T.M. Eads and W.R. Croasmun, *J. Am. Oil Chem. Soc.*, 1988, **65**, 78.
6   S.M. Bociek, S. Ablett and I.T. Norton, *J. Am. Oil Chem. Soc.*, 1985, **62**, 1261.
7   R.H. Atalla and D.L. Vanderhart, *Science*, 1984, **223**, 283.
8   T.J. Foster, S. Ablett, M.C. McCann and M.J. Gidley, *Biopolymers*, 1996, **39**, 51.
9   L. Gambelli, PhD thesis, University of Liverpool, 2001.
10 K. van Putte and J. van den Enden, *J. Am. Oil Chem. Soc.*, 1974, **51**, 316.
11 A.H. Clark and P.J. Lillford, *J. Mag. Res.*, 1980, **41**, 42.
12 K.J. Packer and C.J. Rees, *J. Colloid & Interface Sci.*, 1972, **40**, 206.
13 S. Francis, E.T. Rolls, R. Bowtell, F. McGlone, J. O'Doherty, A. Browning, S. Clare and E. Smith, *Neuroreport*, 1999, **10**, 453.
14 L. Marciani, P. Young, J. Wright, R. Moore, N. Coleman, P.A. Gowland and R.C. Spiller, *Neurogastroenterology and motility*, 2001, **13**, 511.
15 J.C. Lindon, J.K. Nicholson, E. Holmes and J.R. Everett, *Concepts in magnetic resonance*, 2000, **12**, 289.
16 J.C. Lindon, E. Holmes and J.K. Nicholson, *Progress in nuclear magnetic resonance spectroscopy*, 2001, **39**, 1.

SINGLE POINT IMAGING OF BREAD-BASED FOOD PRODUCTS

P. Ramos Cabrer[1], J.P.M. van Duynhoven[2], E.L.A. Blezer[1] and K. Nicolay[3]

[1]Department of Experimental In vivo NMR, University Medical Center, Bolognalaan 50, 3584CJ Utrecht, The Netherlands.
[2]Unilever R&D Vlaardingen. P.O. Box 114, 3130AC Vlaardingen, The Netherlands.
[3]Department of Biomedical Engineering, Eindhoven University of Technology, P.O. Box 513, 5600MB Eindhoven, The Netherlands.

1 INTRODUCTION

Modern food industry is continuously innovating to design and produce products that satisfy consumers' demands (good quality, excellence taste and texture, freshness, ready-to-eat or fast coking, stability during the storage, etc.). For that task, is imperative to use the newest technologies for a better knowledge and control of food properties and behavior. One of the most powerful of these analytical tools is Magnetic Resonance Imaging (MRI), able to provide valuable information about the internal structure and water content and status of food products.

Since bakery represents circa 10% of the overall food industry (products like bread, cakes, cookies, biscuits, snacks, etc. are present on the daily diet of thousands of millions of people) application of MRI in this branch is an important issue. However, the use of MRI is seriously hampered by the nature of bread-like products. Low-water content and/or mobility of bread-like food result in short T2 values, unacceptable for most of the standard MRI techniques. MRI of solid-like food requires the use of one of the available techniques for solids. Among them, we are proposing the use of Single Point Imaging (SPI) methods, because they can be easily implemented on standard MRI instruments. Using SPI methods encoding times of tens of microseconds are feasible, causing an enormous reduction in the T2-weighting in MRI images that largely compensates the characteristic low-proton densities of these products. The minimum achievable encoding time and spatial resolution are basically determined by hardware limitations.

The capability of SPI to image solid materials with short T2 values like polymers, cement, wood, coals or textiles has been already proven, although this technique has been scarcely applied in food science (see reference 1 for a recent review on SPI methods and applications). Here we are presenting SPI images of bread-based products acquired with encoding times as short as 60 μs, showing the potential of this technique in food science.

2 MATERIALS AND METHODS

2.1 Food samples and environmental conditions.

Bread dough was supplied by Unilever R&D Vlaardingen (NL). Dough was stored at −20 °C until its utilization. Dough pieces were thawed for an overnight period in a refrigerator

(4 °C) and proofed at room temperature for 1 h prior to their utilization. All samples were wrapped in cellophane before their placement in the MRI scanner.

Bread, croissants and cookies were purchased from a local store and were used without further preparation. Pieces of 3 cm of the central part of a commercial bread-based snack (1.5 cm salami wrapped by a layer of soft bread) were cut and, where indicated placed in close systems, in which the relative humidity (r.h.) was controlled by saturated solutions of different salts.[2] A ~98% r.h. and a ~43% r.h. environment was achieved using K_2SO_4 and K_2CO_3 respectively (Acros Organics). All experiments were performed at room temperature (20±1 °C).

2.3 MRI experiments.

All experiments were performed on a 4.7 T MRI system (Varian INOVA, Palo Alto, CA, USA) with a 40 cm horizontal bore magnet (Oxford instruments, UK) and a gradient 12 cm set of 22 G/cm. A 16 rods birdcage coil of 6 cm internal diameter operating in quadrature mode was used as transmitter and receiver probe.

2.4 Single point imaging

SPI is a purely phase-encoding technique proposed in 1985 by Emid and Creyghton.[3] In this technique a short non-selective RF pulse is used to excite the whole sample. After an encoding period (tp) one single point of the FID is acquired. Repeating this scheme (shown in figure 1) through n phase encoding steps 1D, 2D or 3D images are obtained. Gradients are switched on before the excitation pulse and maintained constant until the acquisition of the data point. Therefore, rapid switching of gradients needed by other imaging sequences is avoided, reducing ring-down effects and eddy currents.

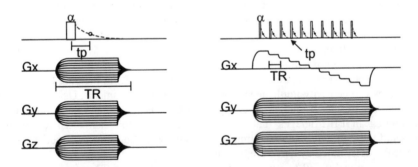

Figure 1 *3D-SPI (left) and 3D-SPRITE (right) pulse sequences. Tp is the encoding time, TR the repetition (recycling) time and α is the excitation pulse angle.*

Since time evolution of the magnetization is not measured (only one point of the FID is recorded), chemical shift, magnetic susceptibility and dipolar and quadrupolar interactions will be not observed.

Although SPI, as described in figure 1, produces T2* weighted images, it's also possible to obtain T1- and T2-weighted images using the appropriate preparation schemes.[4]

SPI traces k-space point by point besides, the use of large gradient amplitudes during long periods of time forces the system to work close to the limits of its duty cycle and resting periods are needed between two consecutive steps in k-space tracing. As a consequence, SPI is a time inefficient technique. Scanning times of several hours are typical for large k spaces, i.e. large fields of view and/or large number of acquired points.

Balcom *et al.* have proposed an improved version of SPI called SPRITE[5] (also shown in figure 1), which reduces scanning times, by ramping gradients in small steps for the first encoding direction. Due to duty cycle restrictions, repetition time (TR) has to be short, introducing certain T1-weighting in the images.

As mentioned above, the shortest encoding time and the highest spatial resolution achievable in SPI technique are both hardware limited. For a given spatial dimension (i), the maximum gradient amplitude needed (G_i) to cover a field of view Δi in *n* steps is,[6]

$$G_i = \frac{n\pi}{\gamma tp \Delta i} \tag{1}$$

According to equation (1), large gradient amplitudes (G) are needed to work with very short encoding times (tp) and/or high resolution (i.e. high $n/\Delta i$ ratio). Scanning time is also indirectly expressed in this equation since, for a given TR, the scanning time will increase with increasing n. Besides, gradient-resting times depend on tp (short encoding times implies large G areas and longer resting times).

3 RESULTS AND DISCUSSION

Since SPI is relatively time inefficient, one must find a compromise between spatial resolution (i.e. high resolution imaging of static systems, suitable for structural analysis) and required time resolution (i.e. dynamical studies with low spatial resolution or small samples). In this paper we are presenting typical examples for these two situations.

3.1 Internal structure of bread and bread like products.

Bread is a complex heterogeneous system where texture (one of the most important quality factors for this product) is determined by the structure of the internal gluten networks. The structural characterization of such complex material should be based on the analysis of high quality images. Most of the published structural analyses of bread structure are based on high quality optical images of sectioned pieces of bread.[7-9] The necessary preparation of the samples (cutting of thin slices) for these techniques could cause damages to the gluten networks, distorting the results. Other published approaches, like the soaking of bread in organic solvents to obtain MRI images,[10-11] lead to good-quality images but, once again, these obtained via an invasive procedure. Here we are presenting high-resolution images of intact bread dough and bread using SPI, without any sample preparation.

Two-dimensional spin-echo (SE) based MRI techniques are suitable for obtaining bread dough images when thick slices are used (figure 2a). However, these images are unsuitable for image analysis because of volume effects. When the slice thickness is reduced to increase the resolution in the slice dimension the signal-to-noise ratio (SNR) drops below acceptable levels (figure 2b). With SPI methods, the acquisition of small voxels is not a problem since the reduction of the signal strength is compensated by a reduction of T2-weighting (figure 2c and 2d). SPI images clearly reveal the structure of the gluten networks of bread dough and parameters like gas cell shape and size can be determined more accurately from these images.

Figure 2 *Bread dough images: (a) Spin-Echo image (3 mm slice, FOV: 7x7 cm, 256x256 points, TE: 9 ms, TR: 3 s). (b) SE image (0.6 mm slice, FOV: 7x7 cm, 256x256 points, TE: 5 ms, TR: 3 s). (c) 0.25 mm slice extracted from (d), 3D-SPI image (FOV: 3x3x5 cm, 120x120x200 points, TE: 300 μs, TR: 0.7 s).*

As s further application of SPI, four different bakery products, i.e. croissant, soft bread, French bread and a cookie, were scanned employing the 3D-SPRITE pulse sequence (tp = 300 μs, FOV = 6.4x6.4x3.2 cm, n = 128x128x64, TR = 5 ms and α = 8°, scanning time = 3.5 h). 2D slices extracted from the central part of the respective 3D data sets are shown in figure 3. These data demonstrate the utility of SPI to obtain high-quality images of bread-like food products

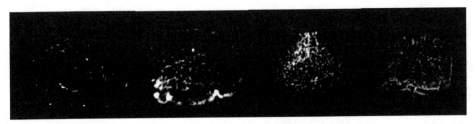

Figure 3 *2D images of a croissant (a), soft bread (b), French bread (c) and a cookie (d), extracted from the respective acquired 3D-SPRITE images.*

Several techniques have previously been presented for the image analysis of sizes and shapes of structures in bread (e.g. gas cells analysis[7-11]). An alternative and interesting approach to describe complex structures of heterogeneous systems is the use of the fractal theory. As an example of the potentiality of SPI for the study of food structure, we applied fractal SPI images of the above bread-based products.

After binarization of the images, respective fractal dimension values (D) were estimated using the box-counting algorithm. D values were calculated from the slope of the logarithmic plot of the number of counted boxes *vs.* box sizes. Values of D=1.719 for croissant, D= 1.724 for soft bread, D=1.758 for French bread and D=1.822 for the cookie were calculated. From the preliminary analysis of these results, it seems that a relationship between softness (a textural property) of bread-like products and a measurable parameter (D) has be pointed out. As softer is the crumb of the product, lower are D values. Since softness is a vague parameter based on sensations thus, further studies are needed to establish a solid relationship between D and other less-subjective parameters.

3.2 Water status of bread-based snacks.

Food properties and stability are mainly determined by water content and distribution. Monitoring of changes in water status caused by (de)hydration and redistribution processes is therefore indispensable in food science. Bulk parameters can be obtained employing NMR spectroscopy, but this information is often insufficient for a complete characterization of heterogeneous systems. MRI, in principle, provides the information in a spatially resolved way, be it that the use of magnetic field gradients limits the information that can be obtained from standard MRI implementations to fractions of water with T2 values above a few milliseconds. SPI techniques, which are able to generate images using encoding times of only tens of microseconds, can offer us much more information about the system, especially if water content and/or mobility is low (like in bread). Since SPI is time inefficient, time course studies with this technique may be limited to low spatial resolution experiments (on standard MRI systems), depending on the minimal time resolution that is required.

As an example of how SPI can deliver valuable information that could not be observed with spin-echo (SE) techniques, we have monitored changes of water content/status in commercial snack. The sample consisted of a 1.5 cm sausage (salami) wrapped by a thin (1 cm) layer of soft bread.

3 sets of 5 images each were acquired using encoding times ranging from 100 μs to 4 ms (SPRITE) and from 4 to 20 ms (SE). Each image was acquired in ca. 5 min (25 min for one set, 1.5 hours for all three sets) using the following parameters; 3D-SPRITE, FOV = 6.4x6.4x3 cm, 64x64x10 matrix, one average, Tp= 100, 200, 300, 500 and 750 μs, TR= 1.5 ms (excitation pulse angle a = 6.5°) for set 1 and Tp= 0.75, 1, 1.75, 2.5 and 4 ms and TR= 5 ms (excitation pulse angle a = 11°) for set 2. SE, three 3-mm consecutive slices, FOV = 6.4x6.4 cm, 64x64 matrix, two average, TE= 4, 6.5, 8, 13 and 20 ms and TR= 2 s.

Every set of images was fitted on a pixel-by-pixel basis to a mono-exponential decay function to generate amplitude maps (from the calculated extrapolated intercepts at TE or Tp=0) and T2 (SE) and T2* (SPI) maps of the snack. Histograms representing T2 distributions were also calculated. Results are shown in figure 4.

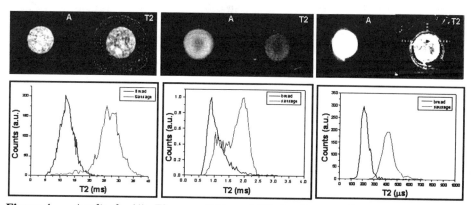

Figure 4   *Amplitude (A), T2 maps (T2) and T2 distributions (bottom row) obtained using SE (left) and SPI with short (center) and ultra-short (right) encoding times.*

Although T2 distributions where obtained for fractions of water with short T2's, the quality of the fittings was poor. For water attached to the macromolecules (short T2's) other fitting models, like a bi-exponential or a gaussian-exponential one, should be applied.

These models were also tried. However the large number of fitting parameters and the fact that only five data points were acquired in each set, leads to a strong correlation between the estimated parameters. Thus, more data points should be acquired to test different fitting models which implies longer acquisition times, that may be too long to follow kinetic processes.

A second experiment was performed with a dual purpose. Firstly, to reinforce the idea that SPI techniques bring to us valuable information about the fraction of water with low mobility, which remains unexposed when only SE techniques are used. Secondly, to gain information on the kinetics of changes in water content and status.

Three SPI images with encoding times tp = 60μs (figure 4a), tp = 500μs (figure 4b) and tp = 1.5ms (figure 4c) and a SE image with TE = 5 ms (figure 4d) were continuously acquired during a 24 hours period (ca. 5 min per image, 20 min for the set of four images). Other acquisition parameters were: for SPI: FOV= 6.4x6.4x3 cm, 64x64x10 points, TR= 650 μs (4a) and TR = 5 ms (4b and 4c), excitation pulse angles: 4° (4a) and 11° (4b and 4c). For SE: FOV= 3 mm slices of 6.4x6.4 cm, 64x64 points and TR= 2 s.

The changes in contrast in the images caused by differences in T2-weighting are evident and specially marked for the bread part of the snack. When very short tp values are used in SPI, plastic parts of the coil and masking tape used to fix the sample (fig. 4a and 4b) are also visible.

Figure 4 *SPI images with Tp = 60 μs **(a)**, 500 μs **(b)** and 1.5 ms **(c)**. Equivalent SE image acquired using TE = 5 ms **(d)**. Plastic parts of the coil and masking tape used to fix the sample are visible in images (a) and (b).*

To get a feeling for the range of possible changes in water status, this experiment was performed in three different atmospheric conditions: (1) room conditions (T=20±1 °C, relative humidity ~70%), hereafter called "normal" environment; (2) closed atmosphere with a controlled relative humidity ~98% (T=20±1 °C), called "wet" environment; and (3) closed atmosphere with a controlled relative humidity ~43% (T=20±1 °C), called "dry" environment.

Plots of the mean pixel intensities (related to water content) of these four sets of images versus time are presented in figure 5 for both bread and sausage in the three mentioned environmental conditions.

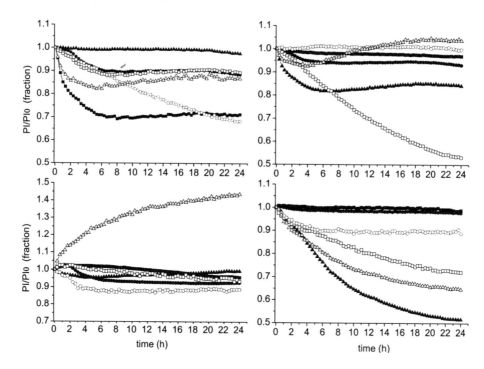

Figure 5 *Bread (hollowed points) and Sausage (filled points) mean pixel intensity variations in the wet (circles), normal (squares) and dry (triangles) environments. Images were acquired with encoding times of 60 μs (top-left), 500 μs (top-right), 1.5 ms (bottom-left) and 5 ms (bottom-right).*

In the snack, palpable dynamic changes have been observed along the studied period and different response of the system to environmental conditions has also been pointed out. Exact composition and storage history of the snack (essential information for a deep analysis of the data) are unknown to us, therefore no conclusions about water transport phenomena in this system will be extracted here. What is important to stress is the fact that data extracted from SPI images (acquired using short encoding times) demonstrate differences on the behavior of water molecules with low mobility (short T2) compared to those of high mobility, reflected on the data extracted from SE images. With the only use of SE experiments (where T2-weighting is strong) this information would remain hidden.

Work is still in progress to optimize the acquisition and data analysis procedures to enable a detailed analysis of the dynamic changes in water status under relevant storage conditions of bread-based products using SPI.

4 CONCLUSIONS

SPI is a promising MRI technique to image food products with natural low-water content (low proton density) and/or mobility (short T2 values).

When scanning time is not an issue (e.g. for structural analysis of bread) SPI provides images with high SNR even when small voxels are acquired (high spatial resolution). This is expected to enable a more detailed structural analysis that possible with spin-echo MRI data

For dynamical studies of water status in food products, SPI provides valuable information of the fraction of water with short and ultra-short T2 values. Since SPI is time inefficient when ultra-short encoding times are required, dynamical studies have to be carried out with reduced spatial resolution.

Acknowledgements

The work described in this paper is funded by Unilever R&D Vlaardingen, with support of the Dutch Ministry of Economic Affairs (Senter).

P.R.C. wishes to thank Professor Dr. José Vázquez Tato (Departamento de Química Física, Universidad de Santiago de Compostela, Spain) who introduced him to the application of fractal analysis in food sciences.

References

1   J. A. Chudek and G. Hunter, *Annu. Rep. NMR Spectrosc.*, 2002, **45**, 151.
2   L. Greenspan, *J. of Research, National Bureau of Standards*, 1977, **81A**, 89
3   S. Emid and J.H.N. Creyghton, *Physica B*, 1985, **128**, 81.
4   D. Beyea, B.J. Balcom, P.J. Prado, A. R. Cross, C. B. Kennedy, R. L. Armstrong and T.W. Bremner, *J. Magn. Reson.*, 1998, **135**, 156.
5   B.J. Balcom, R.P. MacGregor, S.D. Beyea, D.P. Green, R.L. Armstrong and T.W. Bremmer, *J. Mag. res. Ser. A*, 1996, **123**, 131.
6   S. Gravina and D.G. Cory, *J. Mag. Res. Ser. B*, 1994, **104**, 53.
7   K. Kvaal, J.P. Wold, U.G. Indahl, P. Baardseth and T. Naes, *Chemom. Intell. Lab. Syst.*, 1998, **42**, 141.
8   M.C. Zghal, M.G. Scanlon and H.D. Sapirstein, *Cereal Chem.*, 1999, **76**, 734.
9   T. Parkkonen, R. Heinonen and K. Auito, *Food Sci. Technol.*, 1997, **30**, 743.
10   N. Ishida, H. Takano, S. Naito, S. Isobe, K. Uemura, T. Haishi, K. Kose, M. Koizumi and H. Kano, *Mag. Res. Imaging*, 2001, **19**, 867.
11   H. Takano, N. Ishida, N. Koizumi and H. Kano, *J. Food Sci.*, 2002, **67**, 244.

NMR IN FOODS: OPPORTUNITY AND CHALLENGES

Fabiola Cornejo and Pavinee Chinachoti

Department of Food Science, University of Massachusetts, Amherst, MA 01003, USA

1 INTRODUCTION

Over the last few decades, uses of the Nuclear Magnetic Resonance (NMR) to solve food science problems have widen to thanks to rapid development in the NMR techniques. Food scientists have enjoyed the benefits from NMR both from quality control to sophisticated applications. Several literature reviews have been written recently on basic NMR and its wide applications in Food Science. [1,2,3] Molecular behaviors of the food components can be probed by various methods including NMR so that various chemical, physical, and biological properties of food can be explored and exploited. In most simplistic application and thanks to its non-destructive and non-invasive properties, NMR helps us measure the contents of water, oil and other food constituents. NMR sophistication also increasingly meets the utmost challenges in molecular investigation of complex food systems.

Recently, the NMR has been applied in quality control not only to determinate the characteristic of the products such as authentication and detection of adulteration,[7] but also to analyze some characteristic of the product on line such as rheology, bacterial spoilage, etc. Low resolution NMR remains popular for application in measurement of water, such as in carbohydrates. [4,5,6]

NMR can also determinate the oil content as well as authenticity of fats and oils as well as oxidative changes. For example, a high resolution ^{13}C NMR has been used to discriminate between Italian olives oils by cultivars and geographical origins.[7] A study by Isengard and Schmidt[8] showed how low resolution NMR (100 kHz) could be applied to determine the deterioration of the oil during frying by monitoring the polar components.

Applications of NMR cover a wide range in foods, including but not limited to diffusion (e.g., water and oil migration), Magnetic Resonance Imaging (MRI) and micro-MRI, rheo-NMR, structural chemistry, molecular reaction and interaction, mobility, tempering and other process control, microbial activities, and solid-state NMR. This paper will address and highlights some examples or aspects of NMR opportunities and challenges from a food scientist standpoint.

1.1 NMR Investigation of Water in Foods

Physicochemical properties of water can be investigated by molecular relaxation techniques, such as NMR, dielectric, and electron spin resonance. Water determination by NMR is one of the most challenging sciences due to its complex behavior and contributions from various unknown artifacts, which will be highlighted later in the text. Regardless of it complexity, advancement in this area has led to a great number of uses in various fields, such as medical imaging technology, process and quality control, and agricultural applications.

Most foods are comprised of a significant amount of water. Chemical, biological and physical stability of foods no doubt is contributed by the properties of water and how it behaves in the surroundings. Molecular mobility of the water has an important role in the food stability. It affects many physicochemical factors regulating foods biological, chemical and physical quality.[9,10,11,12] Water migration, for instance, is of great interest because most changes in foods are caused by diffusive processes in one form or another. The plasticization effect of water influences many changes in the physical states of foods, such as solubilization, softening due to melting and glassy-to-rubbery transition, and staling. NMR has been used to characterize water mobility in the foods. [13,14,15,16,17]

NMR spectroscopy is based on the measurement of resonant radio frequency adsorption by nuclear spins in presence of an applied magnetic field (B_o). When a magnetic field is applied the nuclei interact with the field and acquire several energy levels. The number of levels is equal to $2I +1$, where I is the spin quantum number. The water has three nuclei that interact with magnetic field, the oxygen-17 (^{17}O), deuterium (^{2}H) and proton (^{1}H) nuclei. Each nuclei is characterized by a spin quantum number (I) therefore each one response different way. The problem is that the chemical exchange and the cross relaxation could affect the proton and deuterium NMR.

Characterization of water of hydration of biological molecules is a controversial on and early attempts to measure mobility of "bound" water in biopolymers and solutes have been subject to criticisms for neglecting unknown contributions of many important phenomena that mostly likely have profound impact on the final interpretation.[3,18,19,20] Multinuclear NMR using proton, deuterium, and oxygen-17 nuclei has been an effective approach to characterize water relaxation and correlation times whose interpretation is complicated by many unknown processes, such as chemical exchange, cross relaxation, and the distribution nature of water populations expected to exist even in a simplest homogeneous system. Applications of more techniques in foods have been possible, such as solid-state, pulse field gradient, imaging, and field-cycling NMR. Unfortunately, this only represents a very small fraction of research work. We have only begun to scratch the surface and more is to be done before NMR fullest potential in food science can be realized.

The majority of studies realized in heterogeneous food measured the proton relaxation in low resolution NMR (in most cases 20 MHz) spectrometer. The NMR spectroscopy has been used to determine not only the water mobility but also the water content. Generally, water is related to the hydrogen atoms in the sample, in other words the total number of protons. Rutledge[21] showed different ways to determinate the water content by time domain – NMR, one with the free induction decay (FID) and the other with pulse sequences (Spin echo and Carr-Purcell-Meiboom-Gill or CPMG). Obtaining the water content by FID signal, after a 90° pulse, is based on the measure of the decay times and the difference between the signal of the protons in the liquid phase and in the solids phase.[21]

1.1.1   Characterizing Water Relaxation Data.   Water molecules can be characterized by the resonance frequency, spin lattice and spin-spin relaxation times. Because some of the water exchange rapidly within the NMR timeframe, signals observed is time averaged. Water molecules that may be associated (chemically or physically) or located a distance away on other domains may be less efficient in exchanging with other water molecules and hence exhibit as a separate population. For the case of proton-proton exchange among domains and molecules also contributes to the relaxation data.

In a non-viscous liquid, $T_1 \sim T_2$ as shown in Figure 1 according to its relatively short correlation time (τ_c). As a material becomes more solid at $\tau_{c < 10^{-6}}$ sec, $T_2 << T_1$.

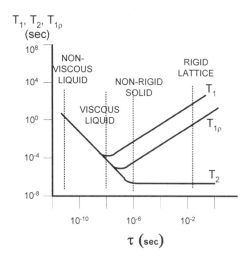

Figure 1   *A typical relationship between T_1, T_2, and $T_{1\rho}$ and τ_c*

A water molecule may experience a number of dynamic processes, reorientation (τ_r), translation or diffusion (τ_D), proton or deuteron exchange with neighboring protons or deuterons (τ_e), and (if associated with a macromolecule) tumbling, anisotropic motion.[22] These describe only the local events.

1.1.2   Extreme-narrowing Limit Systems. In a liquid when an extreme narrowing limit region ($\omega_o \tau_s << 1$ where ω_o is the field strength and τ_s is the slow correlation time of the bound water), relaxation of water has been described by several models assuming with fast-exchange. For example, a model has been used in describing sugar and protein solution assuming fast exchange with two-state anisotropic motion of bound water.[15,23] This assumes that water exists in three populations, free or bulk water (P_w), bound but rapidly reorienting ("fast") water (P^F_{BW}) and anisotropically ("bound") water (P^B_{BW}). This model can be applied to an oxygen-17 NMR data and hence no contribution due to cross relaxation of chemical exchange.[23,24]

Relaxation rate for ^{17}O in a fast exchange regime is a time and population averaged and in an extreme narrowing condition.[25] When a solute is placed in water, it perturbs water molecules creating anisotropic motions on a time scale long compared to the reorientation time of bulk water. A two-site, fast-exchange model taking into consideration of fast and slow motion of bound water[25,26,27], can be used to fit experimental data.[15,23,24]

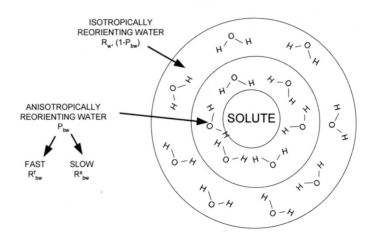

Figure 2   *Water populations according to the anisotropic, two correlation time model for bound water[15,23]*

1.1.3 Heterogeneous System. Vast number of published work on water has been concentrated heavily on hydration of proteins.[28,29,30,31,32,33]

In spatially heterogeneous food systems, diffusive exchange over a distance within the NMR timeframe becomes a significant contributor. This diffusive exchange and cross relaxation have been described elsewhere[3,28,34] and pulse spacing dependence of CPMG (Carr-Purcell Meiboom-Gill) can be used to describe water molecular dynamics based on time and distance scale rather than typical classification according to amount and types of "bound" water[20] to have shown up in heterogeneous systems with protein aggregates, Sephadex gels, and frozen/thawed agarose gels.[35,36,37,38] The distance between regions (each with own characteristic relaxation rates, diffusion coefficients, and resonance frequencies) have to be great enough (e.g., 10 microns) to be observable by NMR; many solid food systems fall in this category. Experiments with various pulse spacing (tau) helps determine the significance of this behavior and additional microscopic or other methods of determination of domain size can be helpful.

One of the methods that helps characterize different water population in complex food systems is T_2 proton relaxation time distribution.[39,40] This method is based in represent the amount of a particular fraction of water molecules in a continuous spectrum. The T_2 distribution has been used to determine the water mobility in raw material and during processing (freezing, drying, gelatinization, etc.). For instance, Tang et. al.[41] applied this method to characterize the distribution of water in native potato starch, considering the granule structure. They also found that the behavior of the water distribution is different from the type A, B, and C starches. This technique was later applied to determine water distribution upon gelatinization and acid hydrolysis[42] and as influenced by harvest time and growth seasons of cassava.[43] Furthermore, due to the non-destructive property, this method could be applied to various real food systems, such as determination of sub-cellular and intracellular redistribution of water in the cellular tissue during processing,[20,44] studies on the water distribution during formation of a curd.[45], and determination of water holding capacity in pork.[46]

Hills and his co-workers have extended their investigation of the proton relaxation to include practical approach to study diffusion and distribution of water in porous medium, such as apple tissue,[47] microbial media,[48], and freezing/freeze drying.[37]

1.1.4 Solid Systems. In a slow-exchange regime, solid state NMR lineshape analysis can provide a first glance of solid and liquid signals. Wide line ^{1}H NMR has been applied to characterized the water mobility in semisolid and solid systems.[49,50,51,52] In high moisture samples, little information can be obtained from ^{1}H NMR, because its spectrum is dominated by the bulk water signal. However, in low moisture samples, the bulk water signal is either absent or greatly reduced and the proton exchange is expected to be much lower.[15, 51] One of the disadvantages of the proton NMR is that it uses is limited by the difficult interpretation of the signal, due to cross relaxation and chemical exchange phenomena.[19,28,53, 54]

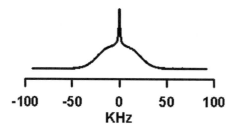

Figure 3 *Wide line ^{1}H NMR spectrum of cellulose at 2.1% water[52]*

Wideline ^{1}H NMR spectrum of a 2.1% water cellulose, obtained from one pulse experiment (90°), is shown in the Figure 3.[52] Narrow narrow (Lorentzian) and wide (Gaussian) components are show indicating some remaining mobility even is such a solid sample, typically considered as glassy. Similar reports are in starch[55] and gum media mixture.[56]

^{2}H NMR can be useful in studying of water. Solid-state ^{2}H NMR of waxy corn[49] as well as other starches (e.g.,potato[43]). Changes in normalized mobile proton signal through 30 to -30 °C range showed an Arrhenius-type relationship for sample with no freezable water (9.3%) but sample with freezable water showed a discontinuity when the temperature passed over an ice formation range.[49] Normalized mobile deuteron intensity of waxy maize starch with no observed DSC freezable water (9.3% water).[49] Activation energy of immobilization of unfreezable water is the heat of fusion for pure water.
Mobile water intensity and time average T_2 for the mobile fraction demonstrates some similarity with water sorption data for gluten (Figure 4) and cellulose.[52] There is a significant implication of water mobility distribution on various stability problems in foods such as microbial activity, diffusion of volatile compounds, gases, water, water soluble components, and reactions such as browning and oxidation. Even in a glassy matrix, small molecular weight components such as water can move around in food shelf-life causing changes. Investigation of mobility sensitivity to temperature change may be quite useful for practical food stabilization technology. Probing water mobility over a temperature and moisture range can provide some useful information with respect to hydration behavior (e.g., Figure 4).

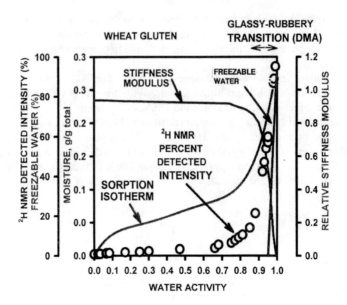

Figure 4 *Water relationship map obtained from NMR, differential scanning calorimetry and dynamic mechanical analysis for vital wheat gluten hydrated with D₂O or H₂O*

1.1.5 Diffusion. PFG –NMR method consists of the standard 90° - τ- 180° - τ- echo sequence, with two identical field gradient pulses of magnitude G, duration δ, and separation Δ.[12] PFG –NMR is an alternative method for measure the translational motion of the molecules, characterized by the diffusion coefficients (D). In contrast to the traditional techniques to measure self diffusion coefficients, NMR can provide data rapidly, and without the need of isotopic substitution.[58,59] In heterogeneous systems the diffusing molecules may encounter a barrier to their motion, a situation called restricted diffusion. In a restricted diffusion, the measured the diffusion coefficient becomes independent of the time (Δ). As a result, the information can provide useful structural information in the range 0.1-10 μm, when the diffusion is restricted on the NMR time scale.[60] By analyzing the PFG - NMR spin echo data, one can obtain information on the geometry of the domains in which restricted diffusion occurs, such information on cell size, droplet size and distribution.[61] The measurement of restricted diffusion has been applied to the study of the water diffusivity in starch water mixture and its relation with the starch gelatinization.[62]

1.1.6 Magnetic Resonance Imaging (MRI). Magnetic Resonance Imaging (NMRI) produces planar or three-dimensional images of an object by mapping spin density and / or spin relaxation times.[12] The fact that MRI provide information of the spatial distribution of the T_1 and T_2 in a system, make it possible to construct a glass transition map (Tg map); with the determination of the T_1 minimum at different temperatures.[63] The MRI also has been applied to determine the flow behavior of fluids (Rheo-NMR),[64,65] water migration during process such as drying, cooking, gelatinization, synerisis etc.,[66,67,68,69,70,71,72] and on-line, quality control.[73]

1.2 NMR of Solids

Cross-relaxation (CR) or proton magnetization transfer,[74,75,76] offers an opportunity to determine solid-liquid protons using a high resolution NMR instrument. CR NMR spectroscopy is a high resolution ^{1}H NMR method for determining the information on the solid component by observing liquid spin system.[74,75] Here, a series of sample irradiation with radio frequency pulse off-resonance (λ) from liquid signal. Dipole-dipole interaction between liquid and solid protons influences the liquid spectrum obtained and with changing λ, a z-spectrum is obtained. This z-spectrum shows a wideline characteristic with combined contribution of mobile and immobile protons (Figure 5).

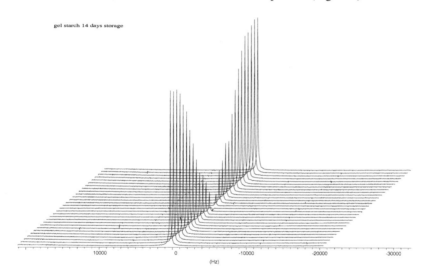

Figure 5   *CR NMR z-spectrua showing ^{1}HNMR spectra for 14days wheat starch gel.[76] Each spectrum was obtained at a given offset frequency in this case ranging over -50 to 50 kHz range.*

1.2.1 Phase Separation and micro- domains-solid state NMR. Much has already been discussed in literature on heterogeneous system involving multi-domains and phase separation. In addition, molecules of same chemical structure in a same crystalline unit cell can be quite different in molecular mobility.[77] Thus, within a domain or phase where a structural miscibility is observed, molecular mobility heterogeneity can be expected. High mobility of a diluent below glass transition temperature, Tg, has been reported, such as in polystyrene,[78] starch,[49,51] and sugar glass.[79]

Carbon-13 cross-polarization (CP) and magic angle spinning (MAS), or ^{13}C CP/MAS, spectra can yield readily improved resolution. Molecular mobility in ms time scale can be characterized using spin-relaxation of both protons and carbons in the laboratory frame and rotating frame ($T_{1\rho}$). Spin locking proton spins in the rotating frame of variable delay (spin lock time) before cross polarization when protons are brought in contact with ^{13}C spins with a specific contact time. Due to dipole-dipole interactions, flip-flops of proton spins or spin diffusion is expected and since the proton magnetization is a function of diffusion coefficient, maximum mean square diffusion distance $<L^2>$ can be used to approximate domain size of heterogeneous polymer blends.[80] When the domain size is small compared with $<L^2>^{1/2}$ or the spin diffusion rate is faster than the relaxation rate, a single average decay constant is observed. Though some polymer blends, though exhibit single Tg, solid-state NMR can reveal non-exponential decay data and in this case a distribution function of $T_{1\rho}$ is applied for multi-phase polymer blends averaging over the $<L^2>^{1/2}$ domain size.

$T_{1\rho}$ can be used to determine specific interactions between polymer moieties in polymer blends.[80,81,82,83] ^{13}C CP/MAS spectra can be analyzed from line shape of carbon-1 to determine polymorphic composition.[84,85] $T_{1\rho}$ for starch, gluten and their mixtures have been reported to exhibit single exponential decay[2,86] dominated by spin diffusion. Bi-exponential decay has also been reported for some cases described as a contribution from diffusion of adsorbed water on different sites on starch.[51] Domain size characterization by this technique is compromised by spin diffusion and an existence of not-so-sharp boundaries between two phases. Exploration using more sophistication may help further understanding heterogeneous-domain systems.

Food structure is of great interest. How micro-domains of amorphous polymers are distributed and contribute to the mechanical and other properties, have been some of the contentious points. Bread, for example, is a complex food composite and understanding molecular origin to bread texture has always involved starch recrystallization but recently the importance of amorphous components have been brought to attention.[87] Solid-state NMR has only started to be utilized to investigate the amorphous components in solids.[2,51,86,88,89] Obviously much more needs to be done to understand the behavior of molecularly unorganized domains which are the majority of most food systems.

It is worth mentioning a method to determine the size and mobility in crystalline and amorphous regions developed by Schmidt-Rohr and his co-workers.[90] This technique applies different of NMR pulse sequences first exciting ^{13}C nuclei with a 90° pulse. A $CP/^{13}CT_1(T_{1,c})$ experiment would follow creating $T_{1,c}$ filter by storing the ^{13}C magnetization along z direction after CP. Goldman-Shen experiment is then applied using an appropriate dephasing time so that the highly mobile phase is selected. In other words, after the dephasing, the crystalline signal vanishes but most of the amorphous signal is retained. This method should work well if the difference in mobility between the amorphous and crystalline regions is adequately large and domain boundaries is relatively sharp. This spin diffusion experiment should give information about the size of the mobile phase. This could still be quite valuable for some circumstances to study semi-crystalline materials.

1.3   Molecular Plasticization

Water acts as a plasticizer in the food polymers, influencing their physical properties. NMR has been used in conjunction with other methods to understand shorter range events of glass transition. Typically, differential scanning calorimetry (DSC) and dynamic mechanical analysis (DMA) are used to measure long-range transition.[91] In vital wheat gluten, ^2H and ^{17}O NMR relaxation was applied in comparison with thermomechanical change during hydration.[92] Figure 4 depicts a summarized view. When oxidized with typical dough conditioning agents (ascorbate and bromate), gluten Tg temperature widened with increasing hydration from 5-20 % water (although mid-points remained unchanged).[93] This corresponded with the increased in water sorption and ^{2}H NMR intensity (line width unchanged). Freezable water from DSC corresponded with ^{17}O data.

Vittadini et al.[56] investigated this relation, studying the mobility of the xanthan and locust bean gum mixtures and its effect in the microbial stability. The results show that, while DSC and DMA produced no evidence of long-ranged glass transition, solid state ^{2}H and ^{1}H NMR strongly indicate a presence of both rigid and mobile components. The relative mobile signal increased with increased moisture content indicative of molecular softening (plasticization). The solid signals observed by NMR were detected up to 25 – 30% moisture content and became insignificant thereafter. The fact that this event could not be monitored with a long range method can only suggests that some plasticization occurs in a much shorter distance and faster time scale. Addition of mannitol (14% dry basis) to the samples caused a decrease in the mobile proton and deuteron signals and corresponding higher survival of the *R japonicum,* found at moisture content of 6%, then it decreases with the increase in mobility (moisture content 6 – 15%). The samples with mannitol had a protective effect on the cells above 17% moisture content (may be related to mobility). Here, microbial response could be related to mobilization on a molecular level where long-ranged glass transition might be detected (According to thermal analyses). In a more heterogeneous system, mold spore germination and growth rate was found to correlate extremely well with T_2 relaxation time both from ^{2}H and ^{17}O NMR experiments.[94]

Because some water closer to the macromolecules may act as a plasticizer, mobility in gluten solids is of some interests. Moreover, molecular process of plasticization is an area not well understood. What location of a molecule that gets intimately mixed with the plasticizer? How does the local (partial) side chain motion affect the backbone motion?

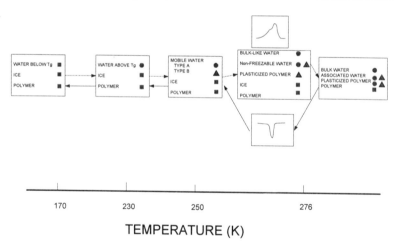

Figure 6   *Example of five thermal equilibrium states in hydrogels*[57]

From a study of water in hydrogels, Quinn et al.[57] described populations and states of water according to various number of thermal equilibrium states as shown in Figure 6.

For food, such characterization is problematic. Combined use of NMR and thermal analyses (and so many others) could provide some useful information. However, to answer specific questions, there is a limitation for studying such complex systems such as foods. One can approach may be to determine the number of water molecules in the first hydration shell but hydration number obtained can vary quite greatly both in proteins[29] and carbohydrates depending on experimental techniques applied (e.g., for sucrose, it can range from 21 based on near infrared spectroscopy to 1.8 based on NMR).[95] Hydrodynamic properties of a macromolecule requires rotation and translational diffusion of molecules. Additional information is also needed in terms of molecular size, shape, and microviscosity. In other words, we need to build more data base on molecular properties of important food molecules. Molecular simulation requires such knowledge. NMR (along with many other techniques) hopefully will increasingly used to determine molecular flexibility and conformational volume a molecule occupies, etc. While these are difficult subjects, NMR can be also applied for practical applications in solving food quality and safety, such as MRI, food structure, and on-line process and off-line quality control.

This paper presents only a small selection of examples of NMR applications in food sciences. In the scope of work presented, NMR has been proven extremely powerful making contribution to the advancements of food science research, teaching, and industrial arena.

Acknowledgements

Authors wish to thank Drs. Thomas Stengle, L. Charlie Dickinson, Brian Hils, and Thomas Eads for their useful discussion over the years. Work was supported in part by Massachusetts Agriculture Experiment Station (MAS 811), US Department of Agriculture, US Army, and the Royal Thai Government (the Ministries of Science and University Affairs).

References

1. S. Divakar, *J. Food Sci. Tech.*, 1998, **35**, 496.
2. A. Gil, P.Belton and B. Hills, 'Applications of NMR to Food Science', in *Annual Reports of NMR Spectroscopy*, ed., G. Webb, Academic Press, San Diego, 1996, Vol. 32, pp. 1-49.
3. P. Belton and I. Colquhoun, *Spectroscopy*, 1989, **4**, 22.
4. W. Kerr and L. Wicker, *Carbohydr Polymers,* 2000, **42**, 133.
5. R. Ruan, Z. Long, A. Song, and P. Chen, *Lebensm.-Wiss. U.-Tech.*, 1998, **31**, 516.
6. P. Cornillon and L. Salim, *Magn. Reson. Imaging*, 2000, **18**, 335
7. F. Ulberth and M. Buchgraber, *Eur. J. Lipid Sci. Techn.*, 2000, **102**, 687.
8. H-D. Isengard and M. Schmidt, *Food Australia,* 2001, **53**, 96.
9. P. Chinachoti, 'New Techniques to Characterize Water in Foods' in *Food Preservations by moisture control: Fundamentals and Applications*, eds., ISOPOW PRACTICUM II, eds., G.V. Barbosa – Canovas and J. Wetti Chanes, Technomic Publishing Company, Lancaster, 1995, pp. 191-207
10. P. Chinachoti and S. R. Schmidt, 'Solute-polymer-water interactions and their manifestations', in: *Water Relationships in Foods*, eds, H. Levine and L. Slade, Plenum Press, New York, 1991, pp. 561-584.
11. L. Slade, and H. Levin, *Crit. Rev. Food Sci. Nutri.*, 1991, **30**, 115.
12. S. Schimidt and H. M. Lai, 'Use of NMR and MRI to study Water Relation in Foods' in *Water Relationships in Food*, eds., S. Levin and L. Slade, Plenum Press, New York, 1991, pp. 405-452.

13. M. Steinberg and H. Leung, 'Some Applications of Wide Line and Pulse NMR in Investigations of Water in Foods' in *Water Relations of Foods*, ed., R. Duckworth, Academic Press, New York, 1975, Chapter 2, pp. 233-248.

14. S. Richardson, I. Baianu and M. Steinberg, *J. Agr. Food Chem.*, 1986, **34**, 17.

15. B. Hills, *Molec. Phys*, 1991, 72, 1099.

16. H. Lai and S. Schmidt, *J. Food Sci.*, 1990, **55**, 1435.

17. S. Richardson and M. Steinberg, 'Applications of Nuclear Magnetic Resonance' in *Water Activity: Theory and Applications to Food*, eds., L. Rockland, L. Beuchat, Marcel Dekker, New York, 1987,Chapter 11, pp. 235-286.

18. J. Finney, J. Goodfellow and P. Poole, 'The structure and dynamics of water in globular proteins' in *Structural Molecular Biology, Methods and Applications*, eds, D. Davies, W. Saenger and S. Danyluk, Plenun Press, New York, 1982, pp. 387 – 426.

19. P. Belton, *Comments Agri. Food Chem.*, 1990, **2**, 179.

20. B. P. Hills, S. F. Takacs and P. S. Belton, *Food Chem.*, 1990, 37, 95.

21. D.N. Rutledge, *Food Control,* 2001 **12**, 437.

22. K. Packer, *Phil. Trans. R. Soc.*, 1977, **278**, 59.

23. P. Belton, S. Ring, R. Botham and B.Hills, *Mol. Phys.*, 1991, **72**, 1123.

24. E. Vittadini, S. J. Schmidt and P. Chinachoti, *Mol. Phys.*, 2001, **99**, 1641.

25. B. Halle, T. Andersson, S. Forsen and B. Lindman, *J. Am. Chem. Soc.*, 1981, **103**, 500.

26. B. Halle and H. Wennestrom, *J. Chem. Phys.*, 1981, **75**, 1928.

27. J. Colquhoun and B. J. Goodfellow, Apectroscopic Techniquies for Food Analysis, ed. By G. C. Levy, American Chemical Society, Washington, DC., 1994 pp 87.

28. B. Hills, *Mol. Phys.*, 1992, **76,** 509.

29. J. G. de la Torre, *Biophys. Chem.*, 2001, **93**, 159

30. G. Schauer, R. Kimmich and W. Nusser, *Biophys. J.*, 1988, **53**, 397

31. A.J. Rowe, *Biophys. Chem.*, 2001, **93**, 93.

32. H. X. Zhou, *Biophys. Chem.*, 2001, **93**, 171.

33. D.J. Winzor, L. Carrington and S. Harding, *Biophys. Chem.*, 2001, **93**, 231.

34. B. P. Hills, S. F. Takacs, and P. S. Belton, *Mol. Phys.*, 1989, 67, 919.

35. B.P. Hills and G. Le Floc'h, *Mol. Phys.*, 1994, **82**, 751.

36. B.P. Hills, K. Wrigh and P.S. Belton, *Mol. Phys.*, 1998, **67**, 193.

37. J. Godward, P. Gunning and B.P. Hills, *Appl. Magn. Reson.*, 1999, **17**, 537.

38. B. Hills and V. Quantin, *Mol. Phys.*, 1993, **79**, 77.

39. B. P. Hills, P.S. Belton and V.M. Quantin, *Mol. Phys.*, 1993, **78**, 893.

40. D. LeBotlan and L. Ouguerram, *Anal. Chim. Acta,* 1997, **349** 339.

41. H. Tang, J. Godward and B. Hills, *Carbohyd. Polym.*, 2000, **43**, 375

42. H. R. Tang, A. Brun and B. Hills, *Carbohyd. Polym.*, 2001, **46**, 7.

43. P. Chatakanonda, P. Chinachoti, K. Srisoth, H. Tang and B, Hills, *Carbohyd. Polym.*, 2002, accepted.

44. B. Hills and B. Remigereau, *Int. J. Food Sci. Tech.*, 1997, **32**, 51.

45. C. Tellier, F. Mariette, J. P. Guillement and P. Marchal., *J. Agr. Food Chem.*, 1993, **41**, 2259.

46. H.C. Bertram, H. J. Andersen, and A. H. Karlsson, *Meat Sci.*, 2001, **57**, 125.

47. B.P. Hills, K. Wright and J. Snaar, *Magn. Reson. Imaging*, 1996, **14**, 715.

48. B. P. Hills, C.E. Manning, Y. Ridge and T Brocklehurst, *Int. J. Food Microbiol.*, 1997, **36**, 187.

49. S. Li, L. C. Dickinson and P. Chinachoti, *J. Agri. Food Chem.*, 1998, **46**, 62.

50. J. Wu, R. G. Bryant and T. Eads, *J. Agri. Food Chem.*, 1992, **40**, 449.
51. S. Tanner, B. Hills and R. Packer, *J. Chem. Soc. Faraday Transactions*, 1991, 87, 2613.
52. E. Vittadini, L. C. Dickinson and P. Chinachoti, *Carbohyd. Polym.*, 2001, **46**, 49.
53. S. Koening, R. Bryant, K. Hallenga and G. Jacob, *Biochemistry*, 1978, **17**, 4348.
54. W. Shirley and R. Bryant, *J. Amer. Chem. Soc.*, 1982, **104**, 2910.
55. Y. Kou, L. C. Dickinson and P. Chinachoti, *J. Agri. Food Chem.*, 2000, **48**, 5489.
56. E. Vittadini, L. C. Dickinson and P. Chinachoti, *Carbohyd. Polym.*, 2002, **49**, 261.
57. F. X. Quinn, E. Kampff, G. Smyth and V. J. McBrierty, *Macromolecules*, 1988, **21**, 3191.
58. T. Norwood, *Chem. Soc. Rev.*, 1994, **23**, 59.
59. H. Watanabe and M. Fukuoka, *Trends Food Sci. Tech.*, 1992, **3**, 211.
60. W. Price, 'Gradient NMR' in *Annual Report on NMR Spectroscopy*, Vol. 32, 1996, edt. Academic Press, pp. 51-142.
61. P. Callaghan, K. Jolley and R. Humphrey, *J. Colloid Interface Sci.*, 1983, **93**, 521.
62. Y. Gomi, M. Fukuoka, T. Mihori and H. Watanabe, *J. Food Eng.*, 1998, **36**, 359.
63. R. Ruan , Z. Long, K. Chang, P. Chen and I. Taub, *Amer. Soc. Agri. Eng.*, 1999, **42**, 1055.
64. J. Götz, W. Kreibich, M. Peciar and H. Buggish, *J. Non Newtonian Fluid Mech.*, 2001, **98**, 117.
65. S. Gibbs, D. Haycock, W. Frith, S. Ablett, and L. Hall, *J. Magn. Reson.*, 1997, **125**, 43.
66. S. Takeuchi, M. Maeda, Y. Gomi, M. Fukuoka and H. Watanabe, *J. Food Eng.*, 1997, **33**, 281.
67. S. Takeuchi, M. Fukuoka, Y. Gomi, M. Maeda, and H. Watanabe, *J. Food Eng.*, 1997, **33**, 181.
68. C. Jenner, Y. Xia, C. Eccles and P. Callaghan, *Nature*, 1988, **336**, 339.
69. H. Song and J.B. Litchfield, *Cereal Chem.*, 1990, **67**, 580.
70. R. Ruan, J.B. Litchfield and S. Eckhoft, *Cereal Chem.*, 1992, **69**, 600.
71. H. Özilgen and R. Kauten, *Process Biochemistry*, 1994, **29**, 373.
72. H. Watanabe, M. Fukuoka, A. Tomiya and T. Mihori, *J. Food Eng.*, 2001, **49**, 1.
73. I. Pykett., IEE transaction on applied superconductivity, 200, **10**, pag.721
74. J. Grad and R. G. Bryant, *J. Magn. Reson.*, 1990, **90**, 1.
75. J. Y. Wu and T. M. Eads, *Carbohyd. Polym.*, 1993, **20**, 51.
76. Y. Vodovotz, L. C. Dickinson, and P. Chinachoti, *J. Agri. Food Chem.*, 2000, **48**, 4948.
77. Y. Wang, P. S. Belton, H. Tang, N. Wellner, S. C. Davis and D. L. Hughes, *J. Chem. Soc.*, 1997, **2**, 899
78. B. J. Cauley, C. Cipriani, K. Ellis, A. K. Roy, A. A. Jones, P. T. Inglefield., B.J. Mackinley and R.P. Kambour, *Macromolecules*, 1991, **24**, 403.
79. B. P. Hills and K. Pardoe, *J. Mol. Liq.*, 1995, **63**, 229.
80. S. Li, L. C. Dickinson, and J. C. W. Chen, *J. Polym Sci.* Part B: Polymer Physics, 1994, **32**, 607
81. J. F. Masson and R. St. J. Manley, *Macromolecules*, 1991, **24**, 5914-5921
82. L. C. Dickinson, P. Morganelli, C. W. Chu, Z. Petrovic, W. J. MacKnight, and J. C. W. Chen. *Macromolecules*, 1988, **21**, 338.
83. J. Schaefer, E. O. Stejskal, and R. Buchdahl. , *Macromolecules*, 1977, **10**, 384.
84. R. P. Veregin, C. A Fyfe, R. H. Marchessault and M. G. Taylor, *Macromolecules*, 1986, **19**, 1030

85. M. Paris, H. Bizot, J. Emery, J. Y. Buzare, and A. Buleon, *Carbohyd. Polym.*, 1999, **39**, 327.

86. S. Li, L. C. Dickinson and P. Chinachoti, *Cereal Chem.*, 1996, **73**, 736

87. P. Chinachoti and Y. Vodovotz, *Bread Staling*, CRC Press, N.Y., 2000.

88. M. Paris, H. Bizot, J. Emery, J. Y. Buzare and A. Buleon, *Int. J. Biol. Macromol.*, 2001, **29**, 127.

89. M. Paris, H. Bizot, J. Emery, J. Y. Buzare and A. Buleon, *Int. J. Biol. Macromol.*, 2001, **29**, 137.

90. WG Hu and K. Schmidt-Rohr, *Polymer*, 2000, **41**, 2979

91. P. Chinachoti, Thermochimica Acta, 1994, **246**, 357.

92. G. Cherian and P. Chinachoti, *Cereal Chem.*, 1996, **73**, 618.

93. G. Cherian and P. Chinachoti, *Cereal Chem.*, 1997, **74**, 312.

94. X. Pham, E. Vittadini, R.E. Levin and P. Chinachoti, *J. Agri. and Food Chem.*, 1999, **47**, 4976

95. S. B. Engelsen, C.Monteiro, C. H. de Penhoat and S. Perez, Biophys. Chem., 2001, **93**, 103.

THE EFFECT OF MICROWAVE HEATING ON POTATO TEXTURE STUDIED WITH MAGNETIC RESONANCE IMAGING

K.P. Nott, S.Md. Shaarani, and L.D. Hall

Herchel Smith Laboratory for Medicinal Chemistry, University of Cambridge School of Clinical Medicine, University Forvie Site. Robinson Way, Cambridge, CB2 2PZ, U.K.

1 INTRODUCTION

Texture is one of the important sensory attributes for consumer acceptance of potato quality.[1] In particular, the loss of moisture and gelatinisation of starch on cooking is a fundamental change by which the potato is perceived to be edible, and these characteristics differ according to heating modality used. In theory, microwave heating is ideal for cooking potatoes because it is selective toward high moisture areas and therefore is likely to be faster and more uniform; in contrast heating in a conventional oven takes far longer due to formation of an insulating layer by drying of the surface. Wilson et al.[2,3] showed evidence for the possible mechanisms associated with cooking potato in either microwave or conventional fan-assisted ovens. They used invasive methods such as alcohol thermometers to measure the temperature, oven drying to measure the moisture distribution, and the uniformity of cooking was evaluated visually by cutting open the potato and locating the greater translucency associated with the gelatinised regions.

Magnetic resonance parameters such as the spin-lattice (T_1) and spin-spin (T_2) relaxation times of water are sensitive to texture of the food matrix[4] because they reflect both the water content and its mobility; thus maps of those parameters provided by Magnetic Resonance Imaging (MRI), non-invasively and non-destructively reflect changes in the moisture distribution and starch gelatinisation[5] on cooking. MRI can also be used to map temperature.[6] At present microwave heating studies have largely been confined to model food gels;[7] although MRI studies of microwave heating of real foods may be complicated by structural changes as well as combined heat and mass transfer processes, each of those processes can be measured by a specific MRI methodology.

This paper is part of a wider study intended to evaluate the applicability of MRI to study microwave processing of real foods. Potato was chosen as a food of contemporary interest which is commonly used in cylinders and slabs as a model porous system for other microwave studies. It is a high moisture content, approximately uniform solid that displays no convection on heating. The objectives of this study were to evaluate the potential of MRI to map separately
- temperature (using phase mapping)
- moisture distribution (using the liquid proton density, M_0)
- starch gelatinisation (using the relaxation times, T_1, T_2)

after microwave cooking a whole potato, with a view in the future to map simultaneously the temperature and moisture profiles in the heated product. Both the heating-rate and -uniformity (affecting starch gelatinisation), and the moisture-loss and -distribution are two interrelated factors that ultimately determine food quality.[8]

2 METHODS

2.1 MRI quantitation

All MRI images were acquired in an Oxford Instruments 2.35 Tesla, 30cm bore magnet (Oxford, U.K.) connected to a Bruker Medzin Technik Biospec II imaging console (Karlsruhe, Germany), using a 14.5cm internal diameter (i.d.) gradient set and a 9.4cm i.d. quadrature bird cage radio frequency (RF) probe, both built in house.

MRI temperature maps were acquired from the phase difference between the heated- and room temperature reference-image, which is linearly calibrated against temperature, using methodology described previously.[7] A scan time of 52 seconds gave a matrix of 128 $\times$ 32 $\times$ 32 voxels.

MRI quantitation of the relaxation times (T_1, T_2) and liquid proton density (M_0) were acquired with 312μm spatial resolution, a 3mm slice thickness and 2 averages. T_1-weighted images were acquired with an echo time (TE) 12ms and repetition times (TR) 0.4, 0.8, 1.6, 3.2 and 7.5s followed by least squares fitting to a mono-exponential saturation recovery model to obtain a T_1 map. A set of 16 T_2-weighted echo images were acquired with TE 12ms and TR 7.5s followed by least squares fitting to a mono-exponential transverse relaxation decay model to obtain a T_2 map; a M_0 map was acquired by extrapolating the fitted transverse decay back to zero time for each pixel.

Bulk T_2 measurements were carried out with TE 1.6ms and TR 10s (256 points); the former was limited by the maximum power that could be applied without destroying the radio frequency probe. Bulk measurements of the Free Induction Decay signal (FID) were carried out with dwell time 20μs and TR 10s.

2.2 Heating experiments

All white potatoes used in this study were bought from Sainsbury's Supermarkets plc. in a 2.5kg bag, from which potatoes between 140 - 145g were selected. Each potato was fixed on a Perspex sheet with a triangular end to allow accurate repositioning in a suitably shaped holder within the probe after heating. Potatoes were cooked for 1 to 5 minutes in a 800W domestic microwave oven (Goldstar, model MA-1164TE) at 100% power with turntable; the oven was pre-warmed by heating 2 litres of water for 5 minutes. The microwave power was determined to be 734W by the IMPI (International Microwave Power Institute) 2 litre test of microwave power output.[9]

2D maps of the relaxation times T_1, T_2 and of M_0 were acquired for the raw and cooked potatoes at 20°C. The inter-sample variation of T_1 and T_2 for raw potatoes was found to be large even for the same variety. Fortunately the non destructive nature of MRI enables each sample to be used as its own reference, thereby eliminating the time consuming survey of different potato varieties, storage conditions, etc; although replicate measurements were made, data from only one set of measurements is given here. Each potato was weighed before and after heating to give the change in weight due to moisture loss.

In a separate experiment, bulk T_2 and Free Induction Decay (FID) measurements were made on the raw and cooked potatoes at 20°C; in this case each potato was also weighed after cooling for comparison against the bulk measurements.

3 RESULTS

Figure 1(a) shows a typical set of temperature maps across a potato after 1 minute heating in a 800W microwave oven at 100% power. Since the MRI temperature mapping methodology requires that the heated and room temperature samples have the same dimensions, temperature maps were difficult to obtain after longer periods of heating because of the resulting shrinkage and distortion of the potato. However those maps did indicate that heat was transferred from the hottest regions to the rest of the potato.

Figure 1(b) shows the percentage weight loss of a potato upon progressive microwave heating for 1 to 5 minutes. A hissing noise was regularly heard between 1 and 1½ minutes indicating the release of steam to reduce the internal pressure after intense heating. This corresponds with the gravimetric measurements which indicate a significant, almost linear, moisture loss after heating for more than 1 minute.

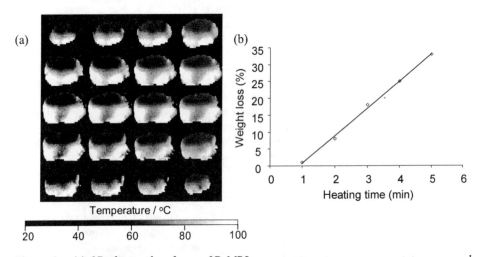

Figure 1   (a) *2D slices taken from a 3D MRI temperature map across a potato measured after 1 minute heating in a 800W microwave oven at 100% power.*
(b) *Gravimetric measurements of potato weight loss (%) versus heating time (minutes) in a 800W microwave oven at 100% power. Cooked weights were measured on the heated potato.*

Figure 2 shows T_2 maps and 1st echo images (TE 12ms, TR 7.5s) for 5 raw potatoes, and the corresponding images after microwave heating for 1 to 5 minutes. Since T_1 maps showed similar trends to those displayed by T_2 they are not shown here; the mean values ranged from *ca* 1330ms for the raw potato to 1240ms after 1 minute and 610ms after 5 minutes heating in the microwave oven. M_0 maps calculated from T_2 fitting, also not shown here, increased with heating time despite loss of moisture indicating that the relaxation decay was not best described by a mono-exponential model; 1st echo images are displayed to provide more representative, yet not quantitative, M_0 maps.

The first observation is that the T_2 maps of the raw potato show sufficient image contrast that it is possible to distinguish at least three different tissue types which are hereafter denoted as:- (a) the pith, (b) parenchyma tissue, and (c) the vascular ring; this image contrast may be a consequence of the different starch content in the three tissues.[10] The T_2 maps clearly demonstrate the loss of this structural differentiation on microwave heating the potato.

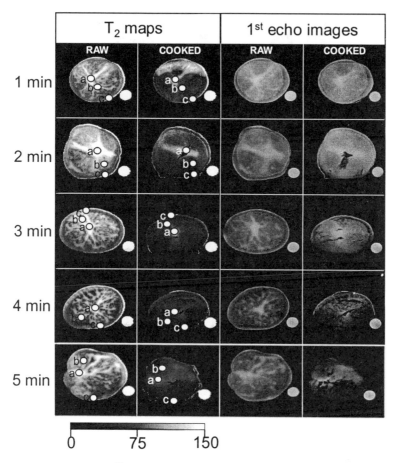

Figure 2 *T_2 maps and 1st echo images (TE 12ms, TR 7.5s) of the raw potatoes and after 1 to 5 minutes heating in a 800W microwave oven at 100% power, acquired at 20°C. Measurements taken from regions of (a) the pith, (b) parenchyma tissue, and (c) vascular ring plotted in Figure 3.*

In agreement with the map of temperature (Figure 1(a)), the T_2 map after 1 minute heating shows that only the bottom of the potato has been cooked to any degree; furthermore it is only after 2 minutes that any significant overall uniformity in texture has been achieved. T_2 measurements taken from the approximate regions shown in Figure 2 for the raw and cooked potato are plotted in the bar chart in Figure 3; convergence of T_2

values for the three different tissues show homogenisation of the potato texture as a whole after 4 minutes. However, further heating leads to moisture loss from the surface which is clearly detected on cutting open the potato, when the surface was noticibly drier.

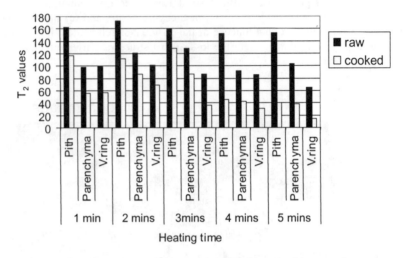

Figure 3   *T_2 values taken from regions representing the (a) pith, (b) parenchyma tissue, and (c) vascular ring displayed in Figure 2.*

Bulk T_2 values measured with a shorter TE (1.6ms) were carried out on the intact potatoes. Figure 4(a) shows a graph of 'weight loss (%)' versus '$-\Delta M_0$ (%)' taken from the sum of the contributions from the T_2 decay. The sum of the M_0 contributions from the different T_2 components should give a representative value of water content which changes linearly with weight loss; this was true for potatoes subjected to 2 to 5 minutes heating ($R^2=0.9993$). However the total M_0 value increased after 1 minute heating indicating an increase in 'MR visible' protons in the cooked compared to raw potato using TE 1.6ms.

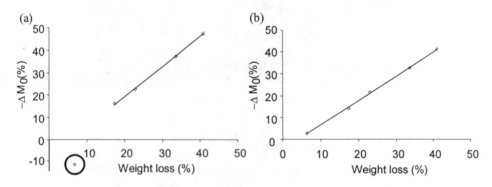

Figure 4   *Weight loss (%) versus $-\Delta M_0$ (%) taken from (a) the sum of the contributions from the bulk T_2 decay, and (b) the first point of the FID. Outlier representing one minute heating is circled in former. After 1 to 5 minutes heating in a 800W microwave oven at 100% power. All measurements were acquired at 20°C.*

Bulk FID measurements were also acquired on the same set of raw and cooked potatoes (dwell time 20μs) to ensure that all the water protons were detected. Figure 4(b) shows a graph of 'weight loss (%)' versus '$-\Delta M_0$ (%)', where M_0 was taken from the first point of the FID. T_2* fitting was deemed inappropriate because the water in a large potato sample in a relatively inhomogeneous magnet would have many different resonance frequencies. Those data gave a good linear fit ($R^2=0.9988$) for all five time points, indicating that all water protons were MR visible. However there was a systematic offset attributable to magnetisation from non water protons in the potato (e.g. starch) as well as from the plastic of the RF probe and holder used for repositioning. Since part of the holder was removed with the potato, we could not subtract their total contribution from the FID.

4 DISCUSSION

The original purpose of this paper was to use MRI to study the mechanism for microwave cooking of potato. Since we are unable to study the process *in situ*, the samples were heated off-line before placing in the magnet for the MRI scan. Consequently each time point had to be measured on a separate potato, using it as its own reference. Inevitably the huge biological diversity between different potatoes, storage conditions etc. effect the inter- and intra-sample variation of the MR properties[11-13] (noticeable in Figure 3); however a detailed analysis of those variations is beyond this preliminary study.

At present the MRI temperature mapping methodology is limited to systems whose overall size and shape do not change with time. Nevertheless, the initial heating patterns for potatoes indicate the location of potential hot and cold spots which could be studied in further detail in real time using fibre optic thermometry within the microwave oven. In this case, there are two localised hot spots at the bottom of the potato which could be a function of a number of variables, including the electromagnetic field in the microwave cavity and potato shape; again those are beyond the scope of the present study.

Wilson et al.[2] postulated that after a short period of 'non uniform' microwave heating of a potato, the internal pressure will increase dramatically due to internal evaporation in the regions of intense heating. Eventually those pressure-induced stresses become too large and cause cellular disruption across the whole potato thereby eliminating resistance to moisture loss. The MR and gravimetric results from this study confirm that hypothesis; thus, between 1 and 2 minutes microwave heating of the potato changes from being localised to being more uniform based on homogenisation of the tissues observed in the raw potato. Three tissue types are distinct in the raw potato because each has different cellular properties; in particular, the abundance and size of starch granules, which are thought to be important for the final texture.[14] Microwave heating causes cellular disruption which results in an even distribution of starch granules, thereby homogenising the texture as observed in the T_2 maps.

Thebaudin et al.[5] showed that the relaxation times T_1 and T_2 of a starch sauce measured at 20°C increased with the previous heating time and temperature. However, the fact that in the case of potato T_1 and T_2 decrease with heating time in the microwave oven despite starch gelatinisation, indicates that there must be significant moisture loss; this was confirmed by gravimetric analysis. However the fact that the calculated liquid proton density, M_0, increases with heating time indicates that the transverse relaxation is not best described by mono- exponential decay sampled with TE 12ms.

Bulk T_2 and FID measurements enable the T_2 and $T_2{*}$ decays respectively to be sampled faster, by shortening the time between the 90° pulse and acquisition of the time domain signal. Using TE 1.6ms, there is an increase in MR visible protons after one minute heating, when part of the potato remains uncooked. It is known that during heating starch goes through three structural changes:- 1) the starch swells by absorbing water and increasing its volume; 2) the crystalline structure melts; and 3) the starch granules are disrupted, releasing amylose.[15] On that basis, the present results indicate that there are 'MR invisible' water protons associated with the starch granule, which are released when the granule is disrupted later in the gelatinisation process. After one minute heating there is a net gain in MR visible protons because the number of water protons released during the gelatinisation process exceeds those lost through moisture loss; however from 2 to 5 minutes, moisture loss prevails causing a net loss.

4 CONCLUSIONS

This study demonstrates both the advantages and difficulties of MRI measurement of combined heat- and mass-transfer in real foods which undergo structural changes, however the technique is sufficiently flexible to enable measurement of the different processes independently. Thus phase mapping gives an MR measurement of temperature that is independent of the moisture content and structural changes, unless they change the magnetic susceptibility. T_2 measurements give a measure of texture, i.e. the distinctive physical structure and composition, either on a macroscopic (tissue) or microscopic (cell) level. Using short acquisition time bulk or imaging MR protocols enable quantitative M_0 measurements to be made of water content and distribution, respectively, during cooking.

Providing that sample preparation and microwave heating conditions are reproducible, it is possible to integrate into a single data matrix, temperature and moisture distribution data for a set of different microwave heating (power, time) and samples (size, shape) obtained using separate MRI scans. Those data sets can then be compared against the models which are often used for process optimisation. However, from a pragmatic viewpoint, those who are responsible for the design, production and testing of real foods packaged for retail domestic distribution, may prefer the direct measurements summarised in this preliminary study. Although those measurement protocols can be applied to most forms of domestic cooking, they are particularly well suited to quantitation of the complex interplay between heat- and mass-transfer which is associated with microwave heating. It is fortunate, though not surprising, that most of the packaging used for foods intended for microwave heating is also suitable for MRI analysis since it is transparent to radio frequency radiation.

Acknowledgements

We gratefully acknowledge funding from the U.K. Biotechnology and Biological Science Research Council (Grant No. D15646), and Dr Herchel Smith for his endowment. SMS would like to thank the Universiti Malaysia Sabah for his scholarship. We also thank Richard Smith, Simon Smith and Cyril Harbird for supply and maintenance of the MRI hardware; Dr Da Xing and Dr Nicholas Herrod for the computer facilities; and Dr Paul Watson for the curve fitting software.

Details of this and other work are available at *www.hslmc.cam.ac.uk*.

References

1. W.G. Burton. *The Potato* (Longman Scientific & Technology, 1989).
2. W.D. Wilson, I.M. MacKinnon and M.C. Jarvis, Transfer of heat and moisture during microwave baking of potatoes. *Journal of the Science of Food and Agriculture*, 2002, **82**, 1070-1073.
3. W.D. Wilson, I.M. MacKinnon, and M.C. Jarvis, Transfer of heat and moisture during oven baking of potatoes. *Journal of the Science of Food and Agriculture*, 2002, **82**, 1074-1079.
4. L.D. Hall, S.D. Evans and K.P. Nott, Measurement of textural changes of food by MRI relaxometry. *Magnetic Resonance Imaging*, 1998, **16**, 485-492.
5. J.Y. Thebaudin, A.C. Lefebvre and A. Davenel, Determination of the cooking rate of starch in industrial sauces: comparison of nuclear magnetic resonance relaxometry and rheological methods. *Sciences des Aliments*, 1998, **18**, 283-291.
6. K.P. Nott and L.D. Hall, Advances in temperature validation of foods. *Trends in Food Science and Technology*, 1999, **10**, 366-374.
7. K.P. Nott, L.D. Hall, J.R. Bows, M. Hale and M.L. Patrick, Three-dimensional MRI mapping of microwave induced heating patterns. *International Journal of Food Science and Technology*, 1999, **34**, 305-315.
8. A.K. Datta and R.C. Anantheswaran, (eds.) *Handbook of Microwave Technology for Food Application* (Marcel Dekker, Inc., New York, 2001).
9. C. Buffler, A guideline for power output measurement of consumer microwave ovens. *Microwave World*, 1991, **10**, 15.
10. G. Lisi´nska and W. Leszczy´nski, *Potato science and technology* (Elsevier Applied Science, London, 1989).
11. A.K. Thybo, I.E. Bechmann, M. Martens and S.B. Engelsen, Prediction of sensory texture of cooked potatoes using uniaxial compression, near infrared spectroscopy and low field H-1 NMR spectroscopy. *Lebensmittel-Wissenschaft Und-Technologie-Food Science and Technology*, 2000, **33**, 103-111.
12. L.G. Thygesen, A.K. Thybo and S.B. Engelsen, Prediction of sensory texture quality of boiled potatoes from low-field H-1 NMR of raw potatoes. The role of chemical constituents. *Lebensmittel-Wissenschaft Und-Technologie-Food Science and Technology*, 2001, **34**, 469-477.
13. H.J. Martens, A.K. Thybo, H.J. Anderson, A.H. Karlsson, S. Donstrup, H. Stødkilde-Jorgensen and M. Martens, Sensory analysis for Magnetic Resonance Image analysis: Using human perception and cognition to segment and assess the interior of potatoes. *Lebensmittel-Wissenschaft Und-Technologie-Food Science and Technology*, 2002, **35**, 70-79.
14. H.J. Martens and A.K. Thybo, An integrated microstructural, sensory and instrumental approach to describe potato texture. *Lebensmittel-Wissenschaft Und-Technologie-Food Science and Technology*, 2000, **33**, 471-482.
15. J. Olkku and C. Rha, Gelatinisation of Starch and Wheat Flour Starch - A Review. *Food Chemistry*, 1978, **3**, 293-317.

APPLICATION AND COMPARISON OF MRI AND X-RAY CT TO FOLLOW SPATIAL AND LONGITUDINAL EVOLUTION OF CORE BREAKDOWN IN PEARS

J. Lammertyn[1], T. Dresselaers[2], P. Van Hecke[2], P. Jancsók[3], M. Wevers[4] and B.M. Nicolaï[1]

[1] Flanders Centre/Laboratory of Postharvest Technology, Catholic University Leuven, Willem de Croylaan 42, 3001 Leuven, Belgium.
[2] Department of Morphology and Medical Imaging, Catholic University Leuven, Herestraat 49, 3000 Leuven, Belgium.
[3] Department of Agro Engineering and –Economics, Catholic University Leuven, Kasteelpark Arenberg 30, 3001 Leuven, Belgium.
[4] Department of Metallurgy and Materials Engineering, Catholic University Leuven, Willem de Croylaan 2, 3001 Leuven, Belgium.

1 INTRODUCTION

To extend the postharvest storage life of fruits and vegetables, controlled atmospheres are commonly applied. Elevated carbon dioxide and decreased oxygen levels during controlled atmosphere storage, in combination with low temperatures, slow down the metabolic activity of these horticultural commodities. Suboptimal or too severe controlled atmosphere conditions can cause irreversible changes in the fruit metabolism resulting in internal storage disorders and large economical harvest losses. Core breakdown in Conference pears is such a storage disorder, which is characterized by brown discoloration of the tissue and development of cavities. To get more insight in the development of the disorder, and more specifically in the spatial and longitudinal evolution, non-destructive repeated measures on the same fruits are necessary. Two non-destructive techniques are frequently referred to in literature for the detection of internal defects in fruits: Magnetic Resonance Imaging (MRI) and X-ray Computer tomography (X-ray CT).

Magnetic Resonance Imaging (MRI) is a non-invasive technique that allows to detect and follow-up the development of storage disorders over time. The study of dynamic characteristics of the aqueous protons (the proton density, the spin-lattice relaxation time and the spin-spin relaxation time) allows non-destructive probing of physiological processes in fruit[1]. MRI has been applied so far for the detection of core breakdown in Bartlett pears[2], the detection of void spaces, worm damage and bruises in fruits[3], quantitative NMR imaging of kiwifruit during growth and ripening[4] and watercore dissipation in apples[5, 6].

X-ray computer tomography, another non-destructive imaging technique, has been explored in the past to study internal disorders like woolliness in nectarines, hollow heart in potato, watercore in apples, spongy tissue in mango. All these disorders affect the fruit density and water content and are, therefore, detectable by means of X-ray measurements[7, 8, 9, 10].

The main objectives of this paper were (i) to assess and compare the suitability of X-ray CT and MRI to detect cavities and brown tissue in pears affected by core breakdown; (ii) to follow both the spatial and time evolution of core breakdown in Conference pears.

2 MATERIALS AND METHODS

2.1 Fruit material

The fruit used for the spatial and time evolution experiments were harvested in a commercial orchard in Zellik (Belgium) and stored under core breakdown disorder inducing conditions (1°C, 10% CO_2 and 1% O_2) until the start of the experiments[11].

2.2 MRI, X-ray CT and digital photographs and image processing

Proton (^{1}H) MR images of whole disordered fruits were collected in a Bruker Biospec (Karlsruhe, Germany) equipped with a horizontal 4.7 T super-conducting magnet with 30 cm bore. Eleven slices were recorded in both the transversal and longitudinal directions (long axis of fruit along longitudinal z-axis of the magnet) using a standard T_1 weighted single-echo (SE) sequence (90° RF pulse), with $TR = 800$ ms and $TE = 18$ ms. The image matrix was 256 x 256 with a field-of-view of 8 cm. The total acquisition time was 3 minutes 27 seconds. The slice thickness was 3 mm and the slice-to-slice distance 5.4 mm, yielding and imaged region of 5.4 cm wide. For the time evolution experiment, 12 pears were followed over a period of 6 months and measured 9 times during this period. MR images at the same settings were taken of a water phantom, which fitted perfectly in the RF coil of the MRI equipment and served as a calibration tool to correct the pear MR images for the coil RF field inhomogeneities. Proton density (N_H) and T_1 maps were obtained from MRI acquired with repetition times equal to 200, 500, 800, 1100, 1500 and 3200ms, respectively[12].

The X-ray CT scans were made on a Microfocus Computer Tomography AEA Tomohawk system (Philips, Netherlands) using a X-ray source (Philips HOMX 161) operated at 53kV and 0.21mA. The exiting radiation was detected by a CCD camera (JAI M50). A HOMX161 object manipulator was used to position the detector and the object table. A Servostep 1700 motor was used to rotate and move the pear during the X-ray CT measurement to suppress CT image ring-artefacts.

Digital photographs were made of thin pear tissue slices, which were cut at the same positions where the X-ray CT and MRI images were taken. Image processing software was written in MATLAB (The Mathworks, USA) to determine the area percentage of affected and healthy tissue, as well as the surface of the cavities and the core[12]. This program was applied to the MRI images, X-ray CT scans and the actual digital photographs.

3 RESULTS AND DISCUSSION

3.1 Comparison X-ray CT, MRI and digital photographs

Figure 1 shows the corresponding transversal tissue slices measured with MRI, X-ray CT and digital photography. On the T_1 weighted MR image (short TR and TE) the tissue in the centre is clearly distinguishable from that at the boundaries of the fruit slice. The

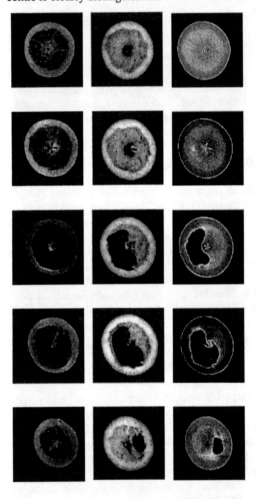

contrast between healthy and affected tissue largely depended on the difference in T_1 value (and proton density) between both types of tissue. Since restricted protons (brown tissue) have a short T_1 value, in a T_1 weighted image the brown tissue should have a high intensity and the healthy tissue, with a higher free water content (long T_1 value) should correspond to a lower intensity. However, as shown on the MR images in Figure 1, the opposite is observed. The brown tissue in the centre has a lower intensity compared to the unaffected tissue at the boundary. This might be explained taking the proton density map into account. The higher the proton density, the higher the pixel intensity on the proton density map. The average proton density values are significantly higher for healthy tissue than for brown tissue. These values (± standard deviation) equal 112,200 ± 9,768 and 58,670 ± 10,871, for unaffected and affected tissue, respectively. On the T_1 weighted images, the brown tissue has a lower pixel intensity than the healthy tissue, resulting from a lower proton density. However, the remaining protons in the brown tissue are highly restricted in motion, what results in a short T_1 value and, hence, a low intensity on the T_1 map. The average T_1 values for healthy and affected tissue were respectively, 1521 ± 202 ms and 1199 ± 192 ms. The T_1 values for brown tissue were significantly (T-test with type I error of 0.05) lower than the T_1 values for unaffected tissue.

Figure 1 *Comparative overview of X-ray CT scans (right), MR images (left) and actual photographs (middle).*

Two processes work in the opposite direction. The brown tissue has a lower proton density resulting in lower signal intensity, but since the remaining protons are highly restricted, the brown tissue has a low T_1 value, which increases the signal intensity on the T_1 weighted images. The decreasing effect of the proton density on the signal intensity of the image exceeds the increasing effect of the T_1 value on the signal, and hence, a net decrease in signal intensity is observed.

In Figure 1 five X-ray CT scans are shown for an affected pear. Cavities are clearly visible as black zones in the slices. Although the contrast is low, the presence of brown tissue is discernable as high pixel intensity areas on the X-ray CT images, which compare well with the visual observations of cut slices. These areas result from lower tissue density and lower water contents. The healthy tissue has a higher density and water content and, therefore, appears as a low intensity area.

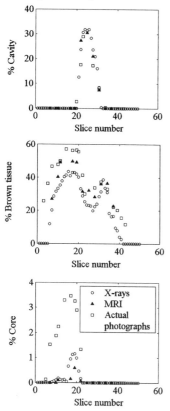

Figure 2   *Comparison of area percentage brown tissue, cavity and core as a function of the slice number for X-ray CT, MRI and photos.*

3.2   Spatial distribution of core breakdown

The algorithm developed to determine the area percentage of core breakdown disorder symptoms was applied to the MR images, X-ray CT scans and the actual digital photographs. In Figure 2 a survey is given of the area percentage affected (brown) tissue, cavity and core as a function of the slice number. Slice 1 and 50 correspond to a transversal slice at the bottom and the top of the pear respectively. A clear relation between the pear size or diameter of the pear and the incidence of the different aspects of the disorder is noticeable. For the X-ray CT scans the percentage brown tissue increases from slice 5 on with increasing pear diameter. At slice 20, the area percentage brown decreases due to a growing cavity. At slice 25 the cavity shrinks again in favour of an increasing percentage of brown tissue. From slice 32 on, no cavity is observed anymore and the percentage of brown tissue decreases with decreasing pear diameter. The total area of the core cavity never exceeds 1.5% of the total slice area. Similar conclusions can be drawn for the MR images. Each MR image, containing information of a 3 mm thick slice, corresponds to a triangle in the plots. Again, the percentage brown increases with the pear diameter and the cavity grows at the expense of brown tissue. This confirms the earlier hypothesis that initially the pears become brown and, subsequently, after a couple of months this brown tissue dehydrates due to moisture diffusion to form a cavity[11, 12].

An obvious similarity was observed between the area percentages of the cavities measured with the three different imaging techniques, performing equally well to determine the cavity area (Figure 2). For the brown tissue, three parallel profiles were present. X-ray CT scans, analysed with the image-processing algorithm, consistently underestimated the percentage of brown tissue on average with 12%. On the MRI images, the percentage brown was higher than for the X-ray CT scans but still on average 5% lower than for the real slices. To explain this parallel shift, a comparative overview of the corresponding X-ray CT, MRI and actual photos is given in Figure 1. In contrast to the real slice photographs, no dark core region could be detected in the X-ray or MRI images (row one of Figure 1). Both techniques showed a region around the core that had a similar intensity as the unaffected tissue, but which was brown as could be observed from the digital photographs of the real slices. This might be attributed to a different measurement principle of the techniques: proton mobility, material density and colour (reflection of visible light).

3.3 Time evolution

3.3.1 Magnetic Resonance Imaging. The time course experiment, in which MR images of pears were taken at regular intervals during storage under disorder inducing conditions, indicated the existence of two different development patterns of core breakdown disorder: a radial and a local pattern.   In the radial pattern, the brown tissue is spread over almost the whole pear slice and its boundaries are parallel to the pear slice boundaries.   A series of MR images of one pear slice, followed during disorder development, is shown in Figure 3. From approximately 58 days after harvest, a ring with a lower intensity starts to develop parallel to the pear slice boundary, indicating the first symptoms of the disorder. The contrast between the affected tissue within this ring and the unaffected tissue outside of it increases with storage time. Opposite to what is expected, the area of brown tissue does not grow during storage from the fruit centre towards the skin, but the biochemical processes causing the tissue breakdown become more pronounced, resulting in lower proton densities and, hence, higher image contrasts with the unaffected tissue.

In the local pattern, the brown tissue appears locally at one or more positions in the pear (Figure 3, right column). Again, 58 days after harvest the first disorder symptoms were noticed on the MR images. Similarly, the area of affected brown did not increase with time, but the contrast with the healthy tissue increased. After 156 days, a cavity was formed in the area of affected tissue. The water content of the affected tissue probably decreased due to moisture transport towards the fruit boundary and resulted in the formation of a cavity. The MR images confirm the earlier stated hypothesis in Lammertyn et al.[11], namely, that cavities are formed at the expense of brown tissue, and that browning starts in the first four months of storage.

A quantitative analysis of the development of the disorder over time was performed following the contrast between healthy and brown tissue over time. The last measured MR image (156 days) was used to define the border between healthy and brown tissue. Subsequently, the average pixel intensity in a tissue region of 10 by 10 pixels at both side of this border was calculated for both types of tissue. The ratio of the average pixel intensity for healthy and brown tissue was used as a measure for contrast, which was then followed in time. The evolution of the contrast value (based on ten measurements on one slice) of 2 randomly chosen pear slices is given in Figure 4. The contrast value is more or less constant during the first two months of storage and increases afterwards, indicating that MRI is able to track the physiological changes taking place in the tissue during the development of the disorder symptoms. The slope of the linear part of the contrast versus time evolution curve was used as a measure for the browning rate. The average browning rate calculated on 10 pears (two slices were randomly chosen per pear) equals 0.0065 ± 0.0005 days^{-1}.

Figure 3 *Time evolution of core breakdown disorder symptoms for a radial (left) and local (right) pattern of browning. Numbers indicated the storage time expressed as days after harvest*

3.3.2 X-ray CT measurements. Simultaneously with the MRI measurements, the time course of core breakdown disorder was studied on a different set of pears, also stored under the same disorder inducing conditions. However, the incipient browning could not be detected with X-ray CT as early as with MRI. This difference in sensitivity between the two tomographic techniques might be attributed to the difference in physical principle on which they are based. X-ray CT measures the attenuation of incident X-rays by the tissue, while MRI measures proton density and mobility. When due to improper controlled atmosphere the cellular membranes disintegrate, and decompartmentation starts at the cellular level, the proton mobility will change, but the density of the tissue will not change much at that time. Later on, when the cell fluids diffuse towards the boundaries of the fruit, the tissue density will change as well, and the internal browning becomes visible at the X-ray CT scans.

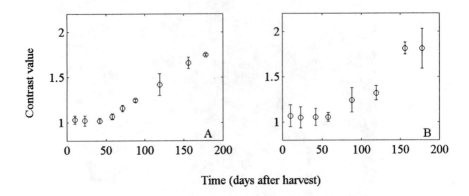

Figure 4   *Average contrast evolution of two randomly chosen pear slices. Vertical bars indicate standard deviations*

4 CONCLUSION

This paper focussed on the application of X-ray CT and MRI as a research tool to get insight in core breakdown disorder in 'Conference' pears. Both tomographic techniques were applied to study the spatial distribution and the time evolution of the disorder symptoms. In the spatial distribution experiment the area percentage affected tissue on the X-ray CT scans and MRI slices were compared qualitatively and quantitatively with each other and with digital photographs of the corresponding actual pear slices. MRI yielded a better contrast between affected and unaffected tissue than X-ray CT scans. In all three techniques the area percentage brown tissue per slice increased with the diameter of the pear, and the contours of the affected tissue were parallel to the pear boundaries, suggesting a relation to gas diffusion properties of tissue and skin. The time course experiment has shown that X-ray CT cannot detect incipient browning as soon as MRI can. However, for severely disordered pears X-ray CT images provided enough contrast to discriminate between affected and unaffected tissue. Differences in MR image contrast between healthy and brown tissue on the MR images allowed to detect incipient browning already after two months of storage. It was found that browning did not grow spatially, but only increased its contrast to healthy tissue during storage. Eventually, cavities grew at the expense of the brown dried tissue. To quantify the browning rate a contrast value was defined based on the ratio of pixel intensities of affected and unaffected tissue.

Acknowledgements

The Belgian Ministry of Small Enterprises, Traders and Agriculture and the Flemish Government are gratefully acknowledged for financial support (project S-5901). This research was also financially supported by EC-FAIR1-CT96-1803 and the Catholic

University Leuven (IDO-project 00/008). Jeroen Lammertyn is Postdoctoral Fellow of the Catholic University Leuven.

References

1. C.J. Clark, J.S. MacFall and R.L. Bieleski, *Sci. Hortic.*, 1998, **73**, 213.
2. C.Y. Wang and P.C.Wang, *HortScience*, 1989, **24**, 106.
3. P. Chen, J.M. McCarthy and R. Kauten, *Trans. ASAE*, 1989, **32**, 1747.
4. C.J. Clark, L.N. Drummond and J.S. MacFall, *J. Sci. Food Agric.*, 1998, **78**, 349.
5. S.Y. Wang, P.C. Wang and M. Faust, *Sci. Hortic.*, 1988, **35**, 227.
6. C.J. Clark and D.M. Burmeister, *HortScience*, 1999, **34**, 915.
7. E.W. Tollner, Y.C. Hung, B.L. Upchurch and S.E. Prussia, *Trans. ASAE*, 1992, **35**, 1921.
8. P. Thomas, S.C. Saxena, R. Chandra, R. Rao and C.R. Bhatia, *J. Hortic. Sci.*, 1993, **68**, 803.
9. T.F. Schatzki, R.P. Haff, R. Young, I. Can, L.C. Le and N. Toyofuku, *Trans. ASAE*, 1997, **40**, 1407.
10. E.G Barcelon, S. Tojo and K. Watanabe, *Trans. ASAE*, 1999, **42**, 435.
11. J. Lammertyn, M. Aerts, B.E. Verlinden, W. Schotmans and B.M. Nicolaï, *Postharvest Biol. Technol.*, 2000, **20**, 25.
12. J. Lammertyn, J., *PhD.- thesis*, Catholic University Leuven, 2002, 210 pages.

THE NMR MOUSE: ITS APPLICATIONS TO FOOD SCIENCE

D.R. Martin[1], S. Ablett[1], H.T. Pedersen[2], and M.J.D. Mallett[3]

[1] Unilever R&D Colworth, Colworth House, Sharnbrook, Bedford, MK44 1LQ, UK.
[2] The Royal Veterinary and Agricultural University, Frederiksberg, Denmark
    Current Address:- Novo Nordisk A/S, Novo Nordisk Park, DK – 2760, Maaloev, Denmark
[3] University of Kent, Canterbury, Kent, CT2 7NR. UK
    Current Address:- Oxford Magnet Technology Ltd., Witney, Oxfordshire, OX29 4BP,UK

1 INTRODUCTION

Low Field (LF) NMR is a widely used technique throughout the food industry. It is a very versatile tool that can measure and quantify a wide variety of parameters, such as solid/liquid fat ratios,[1] droplet sizes in emulsions, and oil and water contents. Information on molecular mobility of different species can also be determined from relaxation time and diffusion measurements.[2] Most LF-NMR or 'benchtop' spectrometers are generally constructed with permanent magnets weighing up to 50 – 100kg hence are not considered portable. They are also designed as small bore instruments with sample tube diameters typically ranging from 10mm to 30mm. Although the technique is often claimed to be non-invasive, it is generally necessary to sub-sample food samples to allow analysis, and therefore the technique cannot always be claimed to be non-destructive. The NMR-MObile Universal Surface Explorer (NMR-MOUSE) is a small compact portable NMR surface scanning device which has the potential to overcome these limitations. The NMR MOUSE could therefore be a truly non-destructive, non-invasive, portable spectroscopic technique.

The NMR-MOUSE was first proposed by Eidmann et al in 1996,[3] and is a LF-NMR device consisting of a one sided magnet and surface mounted RF coil. These replace the magnet and probe in a conventional LF-NMR spectrometer. The one-sided magnet layout allows unrestricted access of large intact samples, hence sub-sampling is no longer required. The magnet is both compact (100 * 50 * 50mm) and light (2 – 3kg) allowing the system to be made portable. The design of the system is such that the magnetic field homogeneity is severely reduced compared to conventional spectrometers. The magnetic field strength decreases rapidly away from the surface of the MOUSE, resulting in a very high magnetic field gradient. The high magnetic field gradients result in the NMR signal decays being dominated by molecular diffusion effects. Previous application of the MOUSE have tended to be limited to fairly solid systems where the effects of molecular diffusion are limited, e.g. polymer materials such as rubber.[4,5]

This paper will outline the general design of the MOUSE, and highlight its potential (and limitations) for food applications. The paper will describe methods of data acquisition applicable to the MOUSE and describe results obtained from oil/water emulsions. A comparison will be made of data recorded on the MOUSE with that from conventional LF-NMR.

2 DESCRIPTION OF THE MOUSE

A picture of the MOUSE is shown in figure 1. This consists of two axially magnetised permanent magnets, placed face to face with anti-parallel magnetisation mounted on an iron yoke. A surface mounted RF coil is located between them. The MOUSE is designed to have a field strength at the surface of about 0.5T, equating to a proton frequency of ~21MHz. The MOUSE is interfaced directly to a conventional LF NMR spectrometer console, in this case a Resonance Instruments MARAN spectrometer (Witney, UK), allowing full control over data acquisition, pulse programming and subsequent data analysis.

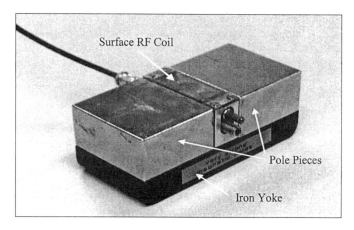

Figure 1   A *picture of the MOUSE.*

A major difference associated with this system compared to conventional LF-NMR is the severely reduced magnetic field homogeneity. Both the static magnetic field (B_0) and RF field (B_1) strength decrease rapidly as a function of distance away from the surface of the MOUSE, resulting in a loss of signal intensity. The decrease in field strength and signal intensity in shown in figure 2. Increasing numbers of inert glass cover slips were inserted between the MOUSE and a standard rubber sample (pencil eraser) while the signal intensity was recorded. The greatest signal intensity was observed with the sample in contact with the surface of the MOUSE operating at a Lamor frequency of 21.3MHz. At this frequency the signal intensity is seen to decrease as the gap between the sample and the MOUSE increases. However since the magnetic field strength decreases rapidly away from the MOUSE surface, retuning the MOUSE to work at lower frequencies should probe further away from the MOUSE surface. At a lower frequency (20.8MHz) the maximum signal intensity is observed some 0.4 mm away from the MOUSE surface. Signal intensity is seen to peak at even lower frequencies even further from the MOUSE surface, albeit at much lower intensity. From this data it is seen that measurements are only possible close to the surface of the MOUSE. It is also estimated that the field gradient is of the order of 20T/m, which is considerably higher than the strength of pulsed field gradients typical found on many LF-NMR spectrometers (< 5 T/m)

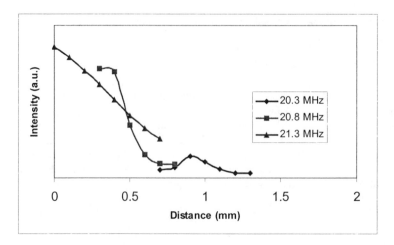

Figure 2   *Plot of signal intensity as a function of sample distance from the MOUSE*
               surface at various Lamor frequencies.

2.1   Single Pulse/Echo Experiments

The presence of the large magnetic field gradient can have an overriding effect on many
conventional NMR experiments. On a conventional LF-NMR spectrometer, the magnetic
field homogeneity is normally sufficiently good that an FID can typically last up to 5ms.
The relaxation behaviour of samples with relatively short spin-spin relaxation times (T_2)
can therefore be determined directly from the FID, or from a solid echo experiment if the
spectrometer dead time is not sufficiently short.

      A solid echo from a standard rubber sample was recorded on a conventional LF-NMR
spectrometer (R.I. MARAN) and is shown in figure 3. The echo can be observed over a
time period of some 5ms. For comparison a solid echo was then recorded for the same
sample on the MOUSE and is also shown in figure 3. In this case the echo is seen to decay
completely within 10μs, which suggests that the intrinsic magnetic field homogeneity of
the MOUSE is of the order of 10μs. This demonstrates that the MOUSE is not suitable for
measuring FID's, since all the signal will decay within 10μs which is roughly equivalent to
the instrument dead time. Similarly the relaxation behaviour of solid samples cannot be
characterised directly from a solid echo experiment, because of the severe magnetic field
inhomogeneity. All measurements on the MOUSE will therefore have to be made with
multiple pulse/echo sequences.

2.2   Multiple Pulse/Echo Experiments

Although the most common multiple pulse sequence for measuring relaxation times on
conventional LF-NMR is the CPMG sequence [90-(τ-180-τ)$_n$], it has been shown for
STRAFI work, which also involves working in very strong field gradients, that [90-(τ-90-
τ)$_n$] echo pulse sequences are equally effective.[7] This can be explained by considering the
theory and behaviour of RF pulses in strong magnetic field gradients, which is described in
detail elsewhere.[6,7,8]

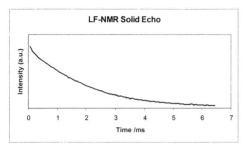
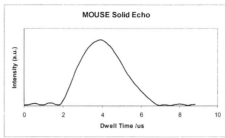

Figure 3   *A solid echo from a rubber sample recorded on a conventional LF-NMR
spectrometer and the MOUSE.*

The frequency spectrum of a rectangular RF pulse is described by a sinc function, the central lobe of which is $2/\tau_p$ wide, where τ_p is the width of the pulse. The slice thickness (Δ_z) excited by this pulse is given by equation 1.

$$\Delta_z = \frac{\sqrt{3}\pi}{\tau_p \gamma G_z} \tag{1}$$

Considering the MOUSE where the field gradient G_z has been estimated at 20Tm^{-1}, it can be shown that a $2\mu s$ rectangular pulse will excite a $50\mu m$ thick slice. Consequently the concept of 'hard' pulses that excite the whole sample evenly, as is the case for conventional LF- NMR, do not exist for the MOUSE. All the applied rectangular pulses on the MOUSE will be effectively slice selective. Since the depth of the sample is invariably much greater than the excitation slice width, the applied RF pulse will also cause some off-resonance excitation leading to a wide distribution of pulses other than 90°. Consequently the amplitude and phase of the magnetisation in the x'-y' plane of the rotating frame will vary across the sample. A more appropriate description of the multipulse echo sequences employed on the MOUSE is given in equation 2, where the net magnetisation tip angle is given by α.

$$[\alpha_x - (\tau - (2)\alpha_x - \tau)_n] \tag{2}$$

Since the NMR sensitive sample volume is reduced to a thin section through the sample, rather than the total sample volume as is normally measured by conventional LF-NMR systems, there is a significant reduction in the signal to noise ratio on the MOUSE compared to conventional LF-NMR.

2.3   Data Analysis

Due to the variation in magnetisation across the sample, the applied RF pulses cause both on- and off- resonance excitation which can generate both direct and stimulated echoes. These echoes can either be constructive or destructive with the result that the initial echo is generally much lower in amplitude than the subsequent echoes.[5] Some typical data recorded on the MOUSE from a multipulse sequence is shown in figure 4 clearly showing the reduced intensity of the first echo. Because of the very fast echo decays rates on the MOUSE, it is generally found to be necessary to record a number of points around the top of each echo to ensure the echo maximum is collected, rather than just digitise a single point, as is the normal case for CPMG measurements on LF-NMR spectrometers.

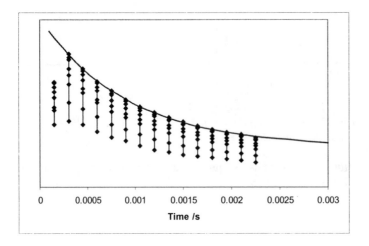

Figure 4   A typical echo train recorded on the MOUSE. Multiple points are collected for each echo.

The overall decay envelope from the second echo onwards does however fit to an exponential decay function. In general the amplitude (A) of an echo is given by equation 3.

$$A_\tau = A_0 \ exp[-(\tau/T_2) - (\gamma^2 G^2 D \tau^3/3)]$$ (3)

For conventional LF-NMR measurements, the homogeneity of the magnet is such that the field gradient G is small, and provided the echo spacing τ is kept sufficiently short, the second term in equation 3 can be neglected. The observed exponential magnetisation decay is then given by equation 4

$$A_\tau = A_0 \ exp[-\tau/T_2]$$ (4)

from which the spin-spin relaxation time, T_2, can be determined directly. In the case of the MOUSE however, the field gradient is so large that the effects of molecular diffusion of liquids within the field gradient cannot be neglected even with small values of τ (50µs), and therefore spin-spin relaxation times cannot be determined directly. The overall decay rate does relate to molecular mobility (with contributions from T_2 and D) and will be referred to as T_2^*. Since the observed magnetisation decay follows an exponential decay function, the initial amplitude of the signal should relate to the amount of sample in the sensitive region and therefore it is considered that the MOUSE should be capable of making quantitative measurements.

3 POTENTIAL OF THE MOUSE FOR TYPICAL FOOD APPLICATIONS.

Thus far it has been proposed that the MOUSE has the potential be used for quantitative relaxation type experiments. Its ability to perform some typical LF-NMR applications in the Food Industry will now be considered.

3.1   Solid Liquid Fat Ratios

One of the most common applications of LF-NMR is the determination of S/L ratios of fats. The NMR signal from fats typically consists of two components, a very rapid decay from the crystalline material and a much slower decay from the liquid. The so called direct method measures the S/L fat ratio directly from the FID. It has already been shown that the MOUSE is not suitable for recording FID's, and therefore the direct S/L fat method cannot be used on the MOUSE. The 'indirect' method involves measuring the amplitude of the liquid oil only at the temperature of interest, and comparing this to the signal amplitude at higher temperatures when all the fat will be liquid. Potentially this method should be possible on the MOUSE, although the lack of temperature control would be a major limitation.

3.2   Molecular Mobility

Conventional LF-NMR is a powerful tool for measuring molecular mobility, both in terms of spin-spin relaxation times (rotational mobility) and diffusion (translational mobility).

3.2.1   Relaxation Times (T_2).   Due to the high field gradients the first term of equation 3 ($-\tau/T_2$) will only dominate the observed magnetisation decay on the MOUSE for solid materials with very slow rates of molecular diffusion, e.g. rubbers and plastics. Potentially the MOUSE could therefore be used to measure relaxation times of these solid materials.

3.2.2   Diffusion.   Due to the effects of molecular diffusion in the high field gradient, the second term of equation 3 ($\gamma^2 G^2 D \tau^3/3$) will dominate for liquids, e.g. water and oil. Potentially the MOUSE could be used to measure rates of diffusion of these materials.

3.3   Droplet Sizing

Droplet sizing is based on diffusion measurements and characterises the restriction to free diffusion within small droplets. Since the MOUSE has the potential to be used for measuring diffusion of liquid samples, it may be possible to develop a droplet size routine.

3.4   Quantification of Oil/Water Contents

Since the magnetisation decay of the echoes generated by the MOUSE has been shown to fit to an exponential decay function, it should be possible to quantify the amount of different molecules present in a sample from the initial intensity of the decay function.

4   EVALUATION OF THE MOUSE

It has been highlighted that for liquid samples the MOUSE has the potential to discriminate different species in terms of their diffusion properties, and quantify the amount of each species present. In order to evaluate the potential of the MOUSE for typical food systems, a series of oil in water emulsion was prepared with varying known oil/water ratios (10% - 70% oil). The aim was to determine how well the MOUSE could discriminate oil from water, and then to establish whether it could be used to measure the oil contents of the emulsions. Results were compared to conventional LF-NMR measurements.

Initially the NMR behaviour of the individual components was investigated. Relaxation times (T_2) of both sunflower oil and water were recorded on a conventional LF-NMR spectrometer (R.I. MARAN) using the standard CPMG pulse sequence. The resulting CPMG decay envelopes are shown in figure 5. As expected the relaxation time of the water ($T_2 \sim 2000$ms) is seen to be considerably longer that that of the oil ($T_2 \sim 150$ms). The large difference in relaxation times allows the two species to be readily discriminated. The NMR behaviour of both species was then recorded on the MOUSE using the $[\alpha_x - (\tau - 2\alpha_x - \tau)_n]$ pulse sequence. The effect of the high field gradient on the MOUSE is to shorten the observed relaxation decay envelopes compared to conventional LF-NMR, due to the effects of molecular diffusion. The resulting decay envelopes for oil and water are also shown in figure 5. In the case of pure sunflower oil the relaxation time is reduced to $T_2^* \sim 100$ms. The effect on water is even more dramatic with the relaxation time reduced to $T_2^* \sim 5$ms! The apparent relaxation rate of water is now much faster than that of oil, but is however sufficiently different from that of oil such that oil and water signals in emulsions should be readily discriminated using the MOUSE.

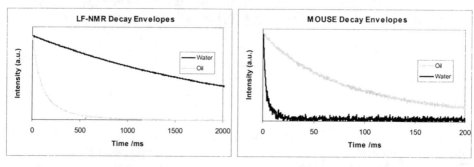

Figure 5   *Decay envelopes for water and oil. Note the much lower signal to noise ratio for the MOUSE data.*

The relaxation behaviour of the emulsions was recorded on both the conventional LF-NMR spectrometer (using the standard CPMG) pulse sequence, and the MOUSE using the $[\alpha_x - (\tau - 2\alpha_x - \tau)_n]$ pulse sequence. All the resulting relaxation decay envelopes were fitted to a sum of two exponential components, from which the relative ratios of the two components were determined. The relative amount of oil in each emulsion, as measured by both conventional LF-NMR and the MOUSE, is given in Table 1 along with the actual concentration of oil used in each emulsions.

Table 1   *% oil content of the emulsions as determined by both LF-NMR and the MOUSE.*

Actual	LF-NMR	MOUSE
10	9	18
19	18	29
30	28	34
39	38	39
48	48	53
58	58	60
67	66	65

It can be seen that there is good agreement between the LF-NMR measured values and the actual oil contents of the emulsions over the complete range of emulsions studied. In the case of the MOUSE data, there is fairly good agreement between the measured values and the actual oil contents for 30 – 70% oil emulsions. However with less than 30% oil present in the emulsion the agreement is less precise.

5 CONCLUSIONS

The MOUSE is a small, compact device which potentially could be made into a portable NMR spectrometer with few restrictions on sample size and geometry. The MOUSE could therefore be made into a truly non-invasive, non destructive spectroscopic technique. The characteristics of the MOUSE are dominated by the presence of a large static magnetic field gradient which is estimated at up to 20Tm^{-1}. The rapid decrease in field strength and sensitivity away from the surface of the MOUSE mean that it is only suitable for making measurements close to the surface of the MOUSE. The NMR signals recorded on the MOUSE are dominated by the high field gradient. The intrinsic magnetic field homogeneity is such that single pulse experiments (i.e. FID's) are not possible and all experiments undertaken must be multiple pulse experiments.

The components of food systems studied by conventional LF-NMR are generally liquid like in behaviour. On the MOUSE the NMR response of these samples will be dominated by the effects of molecular diffusion, such that conventional relaxation time measurements are not possible. It has been shown that the MOUSE is capable of discriminating signals from oil and water in emulsions, primarily due to differences in diffusion, and that the relative ratios of the two components can be quantified. Potentially the MOUSE could also be used for other typical food applications, such as solid liquid fat ratios, diffusion measurements and droplet sizing.

References

1   K. Van Putte and J. Van den Enden, *J. Amer. Oil Chem. Soc.*, 1974, **51**, 316
2   D. Martin, S. Ablett, and M.Izzard, 'Molecular Mobility in Frozen Sugar Solutions' in *Magnetic Resonance in Food Science a View to the Future*, eds. G.A. Webb, P.S. Belton, A.M. Gill, and I. Delagadillo, RSC, 2001 pp.172-178.
3   G. Eidmann, R.Savelsberg, P.Blumler, and B.Blumich, *J. Mag. Res.*, 1996, **A122**, 104.
4   A.E. Sommers, T.J. Bastow, M.I. Burgar, M. Forsyth, and A.J. Hill, *Polym. Degrad. Stabil.*, 2000, **70**, 31.
5   H. Kuhn, M. Klein, A. Wiesmath, D.E. Demco, B. Blumich, J. Kelm, and P.W. Gold, *Mag. Res. Imag*, 2001, **19**, 497.
6   F. Balibanu, K. Hailu, R. Eymael, D.E. Demco, and B. Blumich, *J. Mag. Res.*, 2000, **145**, 246.
7   P.J. McDonald and B. Newling, *Rep. Prog. Phys.*, 1998, **61**, 1441.
8   T.B. Benson and P.J. McDonald, *J. Mag. Res.*, 1995, **A112**, 17.

THE APPLICATION OF NMR IN THE IMPLEMENTATION OF EUROPEAN POLICIES ON CONSUMER PROTECTION

S. Rezzi, C. Guillou, F. Reniero

European Commission, Joint Research Centre, Institute for Health and Consumer Protection, Physical and Chemical Exposure Unit, Ispra (Va), 21020 Italy.

1 INTRODUCTION

High resolution Nuclear Magnetic Resonance (NMR) spectroscopy is a very powerful technique for physico-chemical investigations and for structural determination in particular. NMR is a very powerful quantitative method through the direct proportionality between signal intensity and the number of nuclei in resonance. To date, even when including low resolution NMR, only very few NMR methods have been adopted as official control methods. However, for about a decade, an increasing number of papers have been published on the application of NMR for food studies. Despite this trend, NMR is not yet a common analytical tool used by food control analysts. The reason for this is generally attributed to the cost of purchasing and maintaining NMR equipment. It is also often thought that NMR sensitivity and precision for quantitative determination is not comparable to that attained by other analytical techniques, such as chromatography and mass spectrometry. In the current paper, we do not intend to discuss the above arguments in detail but we would like to express some considerations about the present position and possible further extensions of NMR analysis in the control of food and agricultural products in support of European Policies. In particular, we will discuss some possible applications of NMR analysis in the monitoring of certain aspects of European Regulations regarding traceability, labelling and indication of origin of agricultural and food products.

2 QUANTITATIVE DEUTERIUM NMR

2.1 Generalities

The joint structural and quantitative dimensions of ^{2}H-NMR allows one to observe large variations in deuterium content between the different sites of a given molecule with respect to the expected statistical distribution.[1] Hence, ^{2}H-NMR provides a quantitative tool for establishing the deuterium fingerprint of a food product which can be used as a probe of the chemical, biochemical and technological history of the product.[2,3]

Due to the low natural abundance of deuterium, only monodeuterated molecules are observable. In order to remove the line splittings arising from the scalar couplings between deuterium and proton, the deuterium spectra are always acquired with ^{1}H decoupling. As a consequence, each signal, assigned to each magnetically non-equivalent site, corresponds to one monodeuterated species called an isotopomer. For example, the ^{2}H-NMR spectrum of ethanol exhibits three monodeuterated isotopomers corresponding to the methyl, methylene and hydroxyl sites. ^{2}H-NMR has demonstrated to be very useful for site-specific characterisation of organic molecules, i.e. for the intramolecular distribution of deuterium. Despite its lack of sensitivity, which implies the use of relatively large sample sizes, ^{2}H-NMR provides a "fingerprint" of the deuterium content that can be correlated to its natural or synthetic and/or geographic origin.

The total deuterium content of natural compounds depends on several physical, chemical and/or physiological effects leading to isotope fractionation. These isotope effects can be associated with pedoclimatic parameters, the habitat of the producing plant, the geographic latitude of origin, the deuterium content of local rainwater, the amount of rainfall, evapotranspiration and also with biosynthetic or technological processes.

2.2 EU Wine Databank

In the early 1990's, the European Union adopted the ^{2}H-NMR method for wine analysis as an official method by the Commission Regulation (EEC) No 2676/90, in order to tackle the problem of over-chaptalisation or over-enrichment of wines in Europe. Several publications reported the correlation between the deuterium isotope ratio of wines and their geographical origin.[4,5] Therefore, the application of ^{2}H-NMR as a control method for the wine market needs to take into account the natural variability in Databanks of the wine producing regions, established on a yearly basis. Council Regulation (EEC) No 2048/89 laying down general rules on control in the wine sector, established in Article 16 that the Commission would set up an analytical database for wine sector products at the Joint Research Centre with the aim of co-ordinated and uniform application of the methods of analysis, in particular those based on NMR. The rules for setting up and operating the database were laid down in the Commission Regulation (EEC) No 2348/91. The collection of fresh grape samples for this database was governed by the Commission Regulation (EEC) No 2347/91. In 2000, the Commission Regulation (EEC) No 2729/2000 repealed and replaced the three above mentioned regulations (No 2048/89, No 2347/91 and No 2348/91). From an operational point of view, the Commission established in 1993 the European Office for Wine, Alcohol and Spirit Drinks (BEVABS-Bureau Européen des Vins, Alcool et Boissons Spiritueuses) with the specific objective of co-ordination and management of this European Wine Databank at the Joint Research Centre. Since then, the deuterium NMR data of more than 15000 authentic wines from nine wine producing Member States, have been recorded in the Wine Databank (c.a. 1500 for each vintage). Ten official laboratories from six Member States participate in the NMR measurement of samples. The investment in NMR equipment and the acquisition of know-how is a critical step for the official control laboratories of three wine producing Member States. BEVABS is therefore still carrying out NMR measurements for these Member States while also studying possibilities for helping them in the setting up their own laboratories. Several candidate countries to accession are also wine producing countries and a few of them, in view of harmonising their agricultural policy with that of the European Union, have already set-up their own NMR laboratories for the control of wines. The importance of the wine sector in European agriculture and in the economy of several Member States has certainly enhanced the interest of official bodies for the successful construction of the EU Wine databank and of this European network of official wine NMR laboratories.

2.3 Application to other alcoholic beverages and fermentescible products

Analytical methods very similar to the official European NMR method for wine analysis can be applied for measuring the deuterium content of ethanol from various products. This has been applied to the discrimination of the various botanical and synthetic origins of ethanol or other alcoholic beverages like spirit drinks or beers.[6-9]

Sugars of different botanical origin are characterised by their specific deuterium content. The deuterium content of the non-exchangeable sites of a sugar is correlated to the plant species, affected by different photosynthetic pathways and by the geoclimatic conditions during biosynthesis. For example beetroot sugar shows a lower deuterium content than cane sugar. It has been demonstrated that the deuterium content of a sugar is partly transferred to the ethanol produced by its fermentation.[10-12] The deuterium content of the methyl-group of the ethanol is thus mainly related to that of the fermented sugar. The determination of the addition of sugar to sugar-containing products like musts[13], fruit juices[14-17], honeys[18,19], maple syrup[20], etc. can also be performed by ^{2}H-NMR analysis of the ethanol obtained after controlled fermentation. Similar to wine, it is likely that these methods will also be considered as possible standards (e.g. CEN) or official methods either in Europe or internationally for the control of these kinds of consumer goods. In view of this, and in order to enforce the existing EC legislation, it is essential that official methods used are based on carefully calibrated and well-defined experimental procedures, and that the direct and clear comparison of results is possible both between Community laboratories and internationally. Reference Materials suitable for isotopic analysis, including ^{2}H NMR, of wine, alcohol, fruit juices and sugars have thus been prepared, with the participation of the Joint research Centre (IHCP), throughout the European Commission funded project REFMAT: "Establishing field reference materials for the authentication of food and beverages by isotopic analyses" (DG Research-SMT3-CT96-2086). These materials have now been certified and are available by the European Commission Joint research Centre (IRMM, Institute for Reference Measurements and Materials – Geel, http://irmm.jrc.cec.eu.int).

2.4 Application of ^{2}H-NMR to flavours: the example of raspberry ketone

^{2}H-NMR spectroscopy has been successfully used to provide an isotopic fingerprint characterizing the botanical, semisynthetic or synthetic origin of a wide range of compounds of interest to the flavour industry. The Directives (EEC) No 88/388 of the European Council and No 2000/13 of the European Parliament and of the Council related to flavourings for use in foodstuffs and to foodstuffs labelling, respectively, aim to protect the consumer by requiring detailed labelling which gives the exact nature and characteristics of the food product. More particularly, the term "natural flavouring" may only be used for flavouring substances or flavouring preparations, which are obtained from vegetable or animal materials by appropriate processes (distillation, solvent extraction, enzymatic or microbiological processes). In that context, the precise control of food products is becoming increasingly important since the difference in price between labelled food products (i.e. with a defined geographical origin and/or botanical source) and non-labelled ones is substantial. ^{2}H-NMR has proven to be invaluable in this respect and in the implementation of legislation regarding food products throughout the food industry. As an example, we detail in the following, a ^{2}H-NMR study of raspberry ketone.[21]

Raspberry ketone (RSK) is the main compound responsible for the raspberry aroma, which is widely used in the food industry. Due to the fact that the amount of this compound found in the raspberry fruit itself is very low, an alternative source is necessary to meet demand. ^{2}H-NMR provides a means of distinguishing between "natural" raspberry ketone biogenerated from 4-hydroxybenzalacetone, obtained from para-hydroxybenzaldehyde of extractive botanical origin and acetone produced by sugar fermentation, and other raspberry ketone samples obtained in different "non-natural" ways (Figure 1).

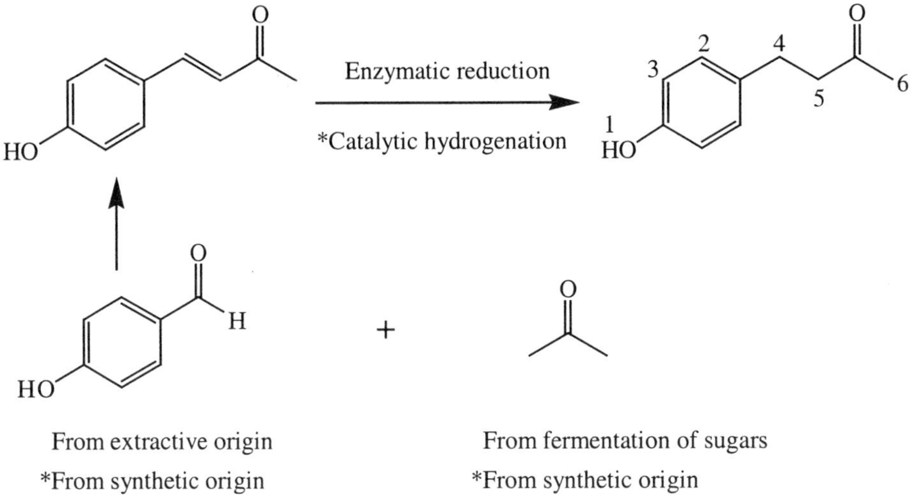

From extractive origin
*From synthetic origin

From fermentation of sugars
*From synthetic origin

Figure 1 *Possible synthesis of RSK (*non-natural)*

We showed that the ^{2}H-NMR measurements, in particular the deuterium pattern of (D/H)$_2$ and (D/H)$_3$ observed for the aromatic ring, allow the clear discrimination between the natural products obtained using biological methods with natural precursors, and those obtained using synthetic precursors (Table 1).

	(D/H)$_2$	(D/H)$_3$	(D/H)$_6$	(D/H)$_3$/(D/H)$_2$
Natural RSK (6 samples)				
Mean	135.4	**170.0**	125.5	**1.26**
SD	11.1	14.4	2.4	0.09
Synthetic RSK (9 samples)				
Mean	128.3	**119.9**	124.8	**0.94**
SD	11.8	8.7	5.0	0.04

Table 1 *^{2}H-NMR study of RSK*

On the other hand, ^{2}H-NMR has been successfully applied to a wide range of flavouring substances such as vanillin[22,23], estragoles and anetholes[22,24], allyl isothiocyanate[25], benzaldehyde[22,26] and p-hydroxybenzaldehyde[27] for some of them.

3   NMR of mixtures for characterisation of food products

3.1 Vegetable oils and fats

Similar to the wine sector, the vegetable oils and fats sector, and more particularly the olive oil sector, are of great economic importance to the European union. In BEVABS we started research on fats and vegetable oil products several years ago using $^{13}C/^{12}C$ and $^{18}O/^{16}O$ ratios for the characterisation of olive oils and for the detection of adulteration. Recently we also showed the possibilities of ^1H-NMR for the detection of the addition of hazelnut oil to olive oil.[28] Further works using this technique are in progress in the shared cost action MEDEO project (EC, DG RTD, Measurements and Testing, contract number: G6RD-CT2000-00440) in which several NMR laboratories are also investigating the possibilities of ^{13}C-NMR to obtain pertinent complementary information useful for the detection of adulteration. Furthermore, our group also participates in another shared cost action, SPREADS project (EC, DG RTD, Measurements and Testing, contract number: G6RD-CT2001-00589), dealing with the determination of the milk fat content of mixed spreadable fats.

3.2 Fish oil and fish products

During the last 10 to 15 years, improvements in methods for aquaculture have lead to a substantial increase in salmon production. This is in line with the growing demand for a high quality product by the consumer. Therefore, the ability to differentiate between wild-caught and farmed salmon is crucial in the fight against fraud in this sector. To date, no reliable method exists to distinguish wild from farmed fish, nor the geographical origin of the fish, leading to real problems with mislabeling and dumping of salmon from non-approved sources. The European project COFAWS (EC, DG RTD, Measurements and Testing, contract number: G6RD-CT-2001-00512) aims to eliminate this risk by developing a validated method for the authentication of the origin of salmon and salmon-based products. In that collaborative work, our group is studying the potential application of high-resolution ^1H-NMR for the analysis of salmon oils. ^1H-NMR is well suited to fish oil analysis allowing the acquisition of quantitative information about total concentration of ω-3 fatty acids, concentration of DHA (22:6n-3), cholesterol, phospholipids, fatty acids methyl esters and mono-, di- and triacylglycerols.[29-31] In this study, we report preliminary results about the application of ^1H-NMR to fish oils extracted from farmed salmon in order to highlight an eventual chemical variability linked to different geographical origins.

We developed a ^1H-NMR procedure in order to obtain a signal to noise ratio and a resolution good enough to allow a correct integration of the signals of interest. In that aim, we selected p-dinitrobenzene as internal standard since its unique signal (8,42ppm) does not interfere with those of the signals of interest. Quantitative spectra were acquired considering the longitudinal relaxation times (T_1) of signals from the sample and from the internal standard previously determined following the inversion-recovery method. Each spectra was obtained in approximately 35 minutes with a fully automated method using an auto sampler, an automatic locking, shimming and processing procedure. Under such conditions, fish oils extracted from the white muscle of 35 salmons (*Salmo salar*) farmed in different north Atlantic locations (Scotland: 3 farms; Norway: 2 farms; and Iceland: 1 farm) were analysed. Figure 2 shows a typical ^1H-NMR spectrum of salmon oil while table 3 indicates the corresponding assignments of the major signals.

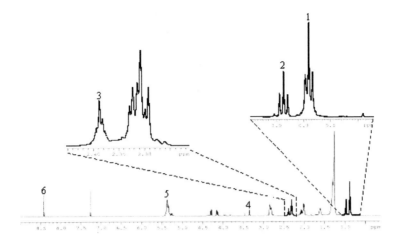

Figure 2 *¹H NMR spectrum of salmon oil*

The areas of signals 1 (methyl protons from all fatty acids, FA, except ω-3 FA), 2 (methyl protons from ω-3 FA), 3 (methylene protons on α and β-carboxylic position of 22:6, Docosahexaenoic acid), 4 (choline methyl protons of phospatidylcholine) and 5 (all olefinic protons) were first standardized with that of the internal standard (signal 6) before being integrated in the data matrix (35 cases x 5 variables) of the Discriminant Analysis in which the sample origin was taken as the grouping variable. Figure 3 shows the graph of the resulting discriminant functions Root 1 and Root 2.

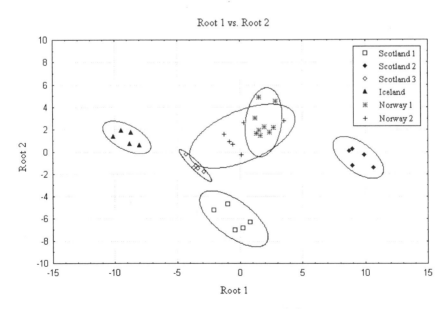

Figure 3 *2D scatterplot of the Discriminant Analysis of ¹H-NMR data from 35 salmon oil samples. The ellipses display the 95% confidence range for each group of salmon oil.*

The salmon oils from the six farms are clearly separated into 5 groups (Scotland 1, Scotland 2, Scotland 3, Iceland and Norway). Samples from the two Norwegian farms seem to be grouped in a single cluster. These preliminary results display the potential of ^{1}H-NMR for salmon oil analysis. Analysis of a greater number of samples, including samples from wild salmon, is in progress. On the other hand, complementary measurements (^{13}C-NMR, ^{2}H-NMR, ^{13}C-IRMS, ^{18}O-IRMS, GC, GC-IRMS) are being carried out by the partners of the COFAWS project.

4 CONCLUSION

NMR technique has demonstrated to be very useful in food analysis. At present, more and more laboratories are applying this technique for food control. It is clear that most of the application, either in ^{2}H-NMR or in the NMR of mixtures very often rely on comparison with data from authentic samples recorded in databanks. For official control purposes, similar to the wine databank, it will be therefore necessary to establish reliable and validated databanks with the help of a network of European laboratories. Obviously, the sufficient comparability of measurements is a prerequisite for both control and the construction of databanks. In the future, the methods of particular interest in the enforcement of EC regulations will need therefore to be validated by collaborative testing. The preparation of ad-hoc reference materials and the development robust data processing will also be necessary.

References

1   G.J Martin and M.L. Martin, *Tetrahedron Lett.*, 1981, **22**, 3525.
2   G.J. Martin, M.L. Martin, F. Mabon and J. Bricout, *J. Am. Chem. Soc.*, **104**, 1982, 2658-2659.
3   M.L Martin and G.J. Martin, *NMR:Basic Prin. Prog.*, 1990, **23**, 1.
4   G.J. Martin, C. Guillou, M.L. Martin, M.T. Cabanis, Y. Tep and J. Aerny, *J. Agric. Food Chem.*, 1988, **36**, 316.
5   A. Monetti, F. Reniero and G. Versini, *Z. Lebensm.-Unters. Forsch.*, 1994, **199**, 311.
6   G.J. Martin, M.L. Martin, F. Mabon and M.J. Michon, *Anal. Chem.*, 1982, **54**, 2380.
7   G.J. Martin, M.L. Martin, F. Mabon and M.J. Michon, *J. Agric. Food Chem.*, 1983, **31**, 311.
8   G.J. Martin, M. Benbernou and F. Lantier, *J. Inst. Of Brew.*, 1985, **91**, 242.
9   J.M. Franconi, N. Naulet and G.J. Martin, *J. Cereal Sci.*, 1989, **10**, 69.
10   G.J. Martin, L.L. Zhang, N. Naulet and M.L. Martin, *J. Am. Chem. Soc.*, 1986, **108**, 5116.
11   G.J. Martin, D. Danho and C. Vallet, *J. Sci. Food Agric.*, 1991, **56**, 419.
12   M.L. Martin, G.J. Martin and C. Guillou, *Mikrochim. Acta [Wien]*, 1991, **II**, 81.
13   A. Monetti, G. Versini, G. Dalpiaz and F. Reniero, *J. Agric. Food Chem.*, 1996, **44**, 2194.
14   G.G. Martin, C. Guillou and M. Caisso, *Flüssiges Obst.* 1991, **1**, 18.
15   G.G. Martin, V. Hanote, M. Lees and Y.L. Martin, *J. AOAC Int.*, 1996, **79**, 62.
16   G.J. Martin, J. Koziet, A. Rossmann and J. Dennis, *Anal. Chim. Acta.*, 1996, **321**, 137.
17   A.M. Pupin, M.J. Dennis, I. Parker, S. Kelly, T. Bigwood and M. C. F. Toledo, *J Agric. Food Chem.*, 1998, **46**, 1369.
18   P. Linder, E. Bermann, B. Gamarnik, *J. Agric. Food Chem.*, 1996, **44**, 139.
19   S. Giraudon, M. Danzart and H. Merle, *J. AOAC Int.*, 2000, **83**, 1401.

20 G.G. Martin, Y.L. Martin, N. Naulet and H.J.D. McManus, *J. Agric. Food Chem.*, 1996, **44**, 3206.

21 G. Fronza, C. Fuganti, C. Guillou, F. Reniero and D. Joulain, *J. Agric. Food Chem.*, 1998, **46**, 248.

22 G. Martin, G. Remaud and G.J. Martin, *Flavour Frag. J.*, 1993, **8**, 97.

23 B. Toulemonde, I. Horman, H. Egli and M. Derbesy, *Helv. Chim. Acta.*, 1983, **66**, 2342.

24 G.J. Martin, M.L. Martin, F. Mabon and J. Bricout, *Sciences des aliments*, 1983, **3**, 147.

25 G.S. Remaud, Y.L. Martin, G.G. Martin, N. Naulet and G.J. Martin, *J. Agric. Food Chem.*, 1997, **45**, 1844.

26 M.L. Hagedorn, *J. Agric. Food Chem.*, 1992, **40**, 634.

27 G.S. Remaud, Y.L. Martin, G.G. Martin and G.J. Martin, *J. Agric. Food Chem.*, 1997, **45**, 859.

28 C. Fauhl, F. Reniero and C. Guillou, *Magn. Reson. Chem.*, 2000, **38**, 436.

29 M. Aursand, J.R. Rainuzzo and H. Grasdalen, *J. A. O. C. S.*, 1993, **70**, 971.

30 T. Igarashi, M. Aursand, Y. Hirata, I.S. Gribbestad, S. Wada and M. Nonaka. *J. A. O. C. S.*, 2000, **77**, 737.

31 R. Sacchi, I. Medina, S.P. Aubourg, F. Addeo and L. Paolillo, *J. A. O. C. S.* 1993, **70**, 225.

CHARGE EFFECTS OF PROTEIN INTERACTIONS AT LIPID-WATER INTERFACE PROBED BY ^{2}H NMR

José A. G. Arêas[1] & Anthony Watts[2]

[1] Dep. de Nutrição, Faculdade de Saúde Pública da Universidade de São Paulo, Av. Dr. Arnaldo, 715, São Paulo, 01246-904, Brazil
[2] Dep. of Biochemistry, Biomembrane Unity, University of Oxford, South Parks Road, Oxford, OX1 3QU, UK

1 INTRODUCTION

Protein is the most important emulsion stabilizer from the Food Science point of view and its original structure and unfolding of the protein at the interface is of paramount importance for emulsion stability. Many different approaches have been taken to gathering more information about the emulsion phenomenon, but most of them result physical, interfacial or even empirical information about this process.[1, 2] Molecular aspects of the lipid-protein interaction at the water-oil interface is scarce and often difficult to obtain. Microscopic, spectroscopic and molecular modelling methods have been used recently and some description at the molecular level has been proposed for emulsion in some systems.[3-7]

In meat products the emulsion formed during the cutting process, where fat globules are surrounded by muscle protein, is further stabilized by gel formation after cooking. The gel backbone is formed by cross-linking of the adsorbed protein and provides the necessary supramolecular support and constitutes the rigid matrix for such products in which texture is achieved.[1, 2] Myosin is an important component of muscle structure. It is a biological motor, responsible for motion in several biological systems.[8-10] Muscle contraction and relaxation, for example, are performed by myosin, powered by the hydrolysis of ATP, moving along actin filaments on a complex supramolecular arrangement of several distinctive protein molecules. In the emulsification process of meat products myosin plays an important role on fat globule stabilization and gel formation after cooking, and it becomes an important component of the network formed.

To gain molecular information about protein-lipid interaction, solid-state ^{2}H NMR that had been successfully used for molecular studies on membrane lipid-protein interactions was employed to describe β-casein-lipid interaction.[3-5] Using this approach this paper reports myosin interaction with the superficial charge of lipid globules and the selective interaction of its tail domain with charged lipids.

2 MATERIAL AND METHODS

Myosin was isolated from bovine lung alveoli and purified by precipitation of an actin-myosin complex.[11] The complex was dissociated by ATP/Mg^{2+} treatment and fractionated with $(NH_4)_2SO_4$ to an 11.3 and 2.2-fold enrichment (measured by Ca^{2+} and K$^+$ EDTA-

ATPase activities). The purity of lung myosin preparation was assessed by PAGE showing a single band of the pure myosin with a molecular mass of ca. 200 KDa. This band presented three types of polypeptide chains when treated with β-mercaptoethanol and SDS, with molecular masses of 176, 19 and 16 KDa, a peptide chain assembly typical for type II myosins.[8-10] Maximum values of enzyme activity were observed at pH 8 and pH 9 for Ca^{2+} and K^+ EDTA-ATPase activities, respectively.

Deuterated DMPC-d$_4$ – 1,2 dimyristoyl-*sn*-glycerol-3-phosphocholine, specifically deuterated at the α- and β- methylenes (DMPC-d$_4$) was prepared from the reaction of CH_3I and 1,2 dimyristoyl-*sn*-glycerol-3-phosphoethanolamine (DMPE-d$_4$), which was synthesized from dimyristoylglycerol with perdeuterated ethanolamine, which, itself, was produced by catalytic exchange of protonated ethanolamine against D_2O.[12-15]

Sample preparation. Samples for NMR experiments were prepared from 50 mg/mL stock solutions of each lipid in $CHCl_3/CH_3OH$ (2:1; v/v) by mixing suitable amounts of each solution to give the desired composition. Solvent was removed in N_2 stream at ambient temperature and dried under high vacuum overnight. The dried lipid or lipid mixture was then fully rehydrated by vortexing in an excess of a buffer containing 20 mM Tris, 20 mM glycine, 1 mM EDTA, 0.6 M KCl, at pH 8.2 (which was the optimum pH and ionic strength for myosin Ca^{2+} ATPase activity) followed by three cycles of freezing under liquid N_2 and thawing at $37^{o}C$. In all cases, deuterium-depleted water (Aldrich, Gillingham, UK) was used.[5]

NMR Measurements. Deuterium NMR spectra were recorded on a Bruker DPX 300 spectrometer. Typical pulse lengths were $4 - 6$ μs for 90° pulses. Solid-state spectra were obtained using a quadrupolar echo pulse sequence ($90^{o} - \tau - 90^{o}$), and for τ a typical value of 30 μs was set. All NMR experiments were recorded with quadrature detection and appropriate phase cycling schemes at 303K (regulated to within $\pm$ 0.5 K by a N_2 temperature control unit). The number of acquisitions varied between 1 and 16K.[16, 17]

3 THE SYSTEM

The deuterated phospholipid employed (DMPC-d$_4$) is shown in Figure 1. Liposomes were made with this phospholipid pure or admixed with charged dimyristoylphosphatidyl-glycerol (DMPG) by three times freeze-thawing the rehydrated phospholipid mixture as described in the experimental section. This resulted in typical 200-400 μm liposomes, checked by scanning electron microscopy, well in the range of fat globules in meat emulsion.[1, 2]

Broad signals (up to 10 KHz) because of the anisotropy of the major quadrupolar interactions are observed in solid-state ^2H NMR spectra of liposomes made of head group deuterated lipids, owing to the restricted amplitude-molecular motion on the liposome surface. The characteristic solid-state double resonance spectra separated by the quadrupole splitting ($\Delta\nu_Q$) of DMPC-d$_4$ in liposomes made of equimolar mixtures of the deuterated lipid and dimyristoylphosphatidylglycerol (DMPG) added of increasing amounts of myosin are presented in Figure 2.

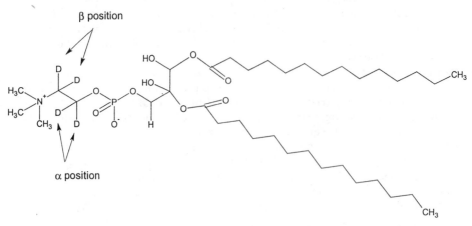

Figure 1   *1,2 dimyristoyl-sn-glycerol-3-phosphocholine (DMPC-d₄) labelled at α and β positions*

4 RESULTS & DISCUSSION

Since the head group conformation is very sensitive to changes in the electrostatic charges of the lipid/water interface, modulation of either structure or electrostatic (or both) properties at the bilayer interface can be monitored directly in the observed ^{2}H NMR spectra of head group deuterated lipids. The character of a non-perturbing probe of deuterium has been used in solid-state NMR experiments for elucidating, at the molecular level, lipid-protein interactions by using deuterated head group lipids in biological [18-20] and food systems. [21-23]

NMR spectra of these liposomes result typical deuterium solid-state exhibiting double peaks for the α and β deuterons unequivocally assigned (Figure 2). Changes in the quadrupole splittings ($\Delta\upsilon_Q$) were used to monitor the electrical charge changes at the surface of the liposomes. Previous experiments have shown that myosin-lipid interaction was negligible in neutral liposomes made only of DMPC-d₄.[24] Addition of the charged phospholipid DMPG altered the quadrupole splittings according to the molar ratio of the charged and uncharged phospholipid. A decrease in $\Delta\upsilon_Q$ of the deuterons at the α and an increase in the ones at the β position are typically observed with increase in DMPC-d₄ molar fraction.[25] This was used to build a "calibration curve" relating α and β deuterons quadrupole splittings to the amount of DMPG present in the liposomes as shown in Figure 3. The calibration curve shown in Figure 3 was built with liposomes made of equimolar mixtures of both lipids (molar fraction of both = 0.5); DMPC-d₄ concentration in the mixture that was then increased up to a 2:1 DMPC-d₄/DMPG molar ratio (DMPC-d₄ molar fraction = 0.75).

Type II myosins present the same typical molecular assemble with a globular domain that presents ATPase activity and sites for specific interaction with actin in the muscle contraction/relaxation process. Another characteristic domain is a long tail made of a α helix structure with a sequence such that hydrophobic amino acid residues repeat at one side of the helix and lysine is present at regular intervals (usually each 28 amino acid residues) at the other side. This determines that type II myosin molecules are able to interact to each other repeatedly through the hydrophobic residues at the tail with the charged lysine residues of this domain facing the water milieu.[8-10]

The strong interaction of myosin and liposomes made of mixtures of DMPC-d$_4$ and DMPG observed by solid-state ^{2}H NMR probably occurs between the positively charged lysine residues of the protein molecule and the negatively charged DMPG component of the liposomes. Thus, increasing amounts of protein interacting with these liposomes would promote a gradual neutralization of their surface electrical charge resulting a sort of lateral phase separation of the deuterated lipid as more DMPG interacts with the protein.

An experiment was performed with addition of increasing concentrations of myosin to liposomes made of equimolar mixtures of DMPC-d$_4$ and DMPG. The quadrupolar splitting of α and β deuterons were then determined from ^{2}H solid-state spectra recorded for each myosin concentration added (Figure 2). The calibration curve presented in Figure 3 was then used to calculate the DMPC-d$_4$ molar fraction as protein/lipid molar ratio increased. The results obtained for this calculation using either α or β deuterons were equivalent and are presented in Figure 4.

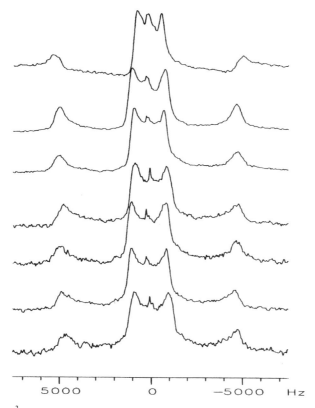

Figure 2 ^{2}H NMR spectra of DMPC-d$_4$/DMPG liposomes (40 mg of lipid) at 1:1 molar ratio, after myosin binding at L/P ratios at 303 K. ($\Delta\upsilon_Q$ of α and β deuterons used to calculate the amount of free DMPC from standard curve in Figure 3)

The calculated molar fraction of deuterated lipid in liposomes made of equimolar mixture of both lipids as myosin concentration increased, presented in Figure 4, indicates that more DMPC-d_4 is freed up to a maximum as protein concentration increases. These results are an indication that the interaction occurs through the positively charged lysine residues and the negatively charged DMPG.

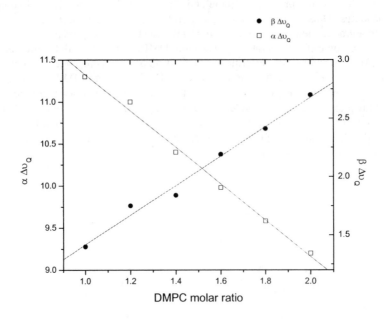

Figure 3   *Effect of composition in the quadrupole splittings of α and β for DMPC deuterons as a function of the molar ratio DMPC/DMPG*

The results observed in this work indicates a supramolecular arrangement for myosin/lipid interaction such that the myosin "tail" lays on lipid globule surface whereas the "head" is up and able to interact with other proteins. In the gel formation that takes place during the cooking process of meat emulsions this arrangement would favour disulfide bridges formation as cystein residues are present only on the head domain of myosin. Microscopy of myosin gel formation indicates that interaction may occur predominantly through the globular head domain.[7] The use of low molecular weight emulsifiers that usually present negative charge may favour myosin interaction with fat in meat emulsions improving performance and final quality of the product.

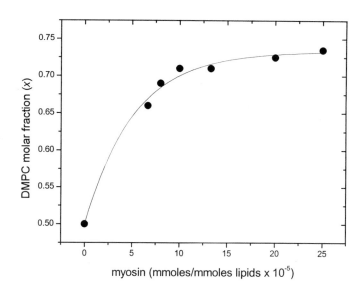

Figure 4 *DMPC molar fraction (x) in the free lipid bilayer as a function of myosin added*

References

1 M.A. Bos and T.vanVliet, *Adv. Colloid Interface Sci.*, 2001, **91**, 437.
2 S.E. Hill 'Emulsions' in Methods of Testing Protein Functionality, ed. G.H. Hall, Blackie Academic & Professional, London, 1996, chapter 6, pp. 163-185.
3 M.M. Cassiano and J.A.G. Arêas, *J. Mol. Structure* (THEOCHEM), 2001, **539**, 279.
4 J.A.G. Arêas, M.M. Cassiano, C. Glaubitz, G.J. Gröbner and A. Watts, 'Interaction of β-casein at an emulsion interface studied by ^{2}H NMR and molecular modeling' in *Magnetic Resonance in Food Science: A View to the Future*, eds. G.A. Webb, P.S. Belton, A.M. Gil and I. Delgadillo, Royal Society of Chemistry, Cambridge, 2001, pp 193-201.
5 J.A.G. Arêas, G. J. Gröbner, L.B. Pellacani, C. Glaubitz and A. Watts, *Magn. Res. Chem.*, 1997, **35**, S119.
6 J.A.G. Arêas, G.J. Gröbner, C. Glaubitz, and A. Watts, *Biochemistry*, 1998, **37**, 5582.
7 A.M. Hermansson, 'Microstructure of protein related to functionality', in *Protein Structure-Function Relationship*, eds. R.Y. Yada, R.L. Jackman, and J.L. Smith, Blackie Academic & Professional, Glasgow, 1994, pp. 22-42.
8 R.J. Adams, and T.D. Pollard, *Nature,* 1986, **322**, 754.
9 T.D. Pollard, S.K. Doberstein and H.G. Zot, *Annu. Rev. Physiol.*, 1991, **53**, 653.
10 R.E., Cheney and M.S. Mooseker, *Curr. Opin. Cell Biol.*, 1992, **4**, 27.
11 D.W. Frederikssen and D.D. Rees, *Meth. Enzymol.*, 1982, **83**, 292.
12 H. Eibl, *Proc. Natl. Acad. Sci.*, 1978, **75**, 4074.

13 M.G. Taylor and I.C.P. Smith, *Chem. Phys. Lipids,* 1981, **28**, 119.

14 K. Harlos and H. Eibl, *Biochemistry* 1980, **19**, 895.

15 F. Sixl and A. Watts, *Proc. Natl. Acad. Sci.,* 1983, **80**, 1613.

16 J.G. Powles and J.H. Strange, *Proc. Phys. Soc.* 1963, **82**, 6.

17 K.R. Jeffrey, *Bull. Magn. Reson.* 1981, **3**, 69.

18 M.R. Morrow, S. Taneva, G.A. Simatos, L.A. Allwood and K.M.W. Keough, *Biochemistry* 1993, **32**, 11338.

19 A. Watts, *J. Bioenerg. Biomembr.* 1987, **19**, 625.

20 *Biochemistry,* 1998, **37**, 5587.

21 P.G. Scherer and J. Seelig, *Biochemistry* 1989, **28**, 7720.

22 F.M. Marassi, S. Djukic and P.M. Macdonald, *Biochim. Biophys. Acta* 1993, **1146**, 219.

23 M. Roux, J.M. Neumann, R.S. Hodges, P.F. Devaux and M. Bloom, *Biochemistry* 1989, **28**, 2313.

24 J.A.G. Arêas and A. Watts, *Biochem. Soc. Trans,* 1993, **21**, 80S.

25 B.B. Bonev, W.C. Chang,

THE CLASSICAL HAUSDORF MOMENTUM PROBLEM APPLIED TO THE LRP-NMR MEASUREMENTS; STABLE RECONSTRUCTION OF THE T_2 DISTRIBUTION AND MAGNETIC SUSCEPTIBILITY DIFFERENCE DISTRIBUTION

V.M. Cimpoiasu and R. Scorei

Biochemistry Laboratory, University of Craiova, A.I.Cuza 13, Craiova, 1100, Romania

1 INTRODUCTION

In the LRP NMR measurements, when we deal with complex systems, because of low S/N ratio, the quantitative estimation of relaxation times (true number of exponentials) and their corresponding populations of protons is a real problem because of the instabilities of solutions and depending on the experimental technique and observations. Therefore, in order to eliminate these difficulties and according to the time domain NMR technique, it is necessary to develop a new method to analyse of such complex systems [1]. We show that besides of extraction of distribution function of relaxation times with good level of confidence some parameters like pondered mean of relaxation times "weighted T_2" T_{2W}, stable at noise, is useful also for investigation of such complex systems.

For measurements of self diffusion in cellular, porous, liquids mixture, etc, with CPMG sequence, in presence of steady external magnetic field gradient, the diffusion of the spins (protons) through local inhomogeneities (developed near the boundaries) leads to a lost of spins coherence and thus influence the nuclear spin relaxation [2]. In such systems, the sink of relaxation at boundaries may be due to the presence of paramagnetic centres at the solid surface, to the momentary reduction of the rotation trembling of molecule at surface and to the dephasing caused by strong local magnetic field gradients. For heterogeneous systems, the non-homogeneity of susceptibility magnetic field may be considered like a spatial interaction .modulated by diffusivity of molecule. Thus, the spin diffusion in susceptibility variation of local magnetic field leads to a slow modulation and influences T_2[3]. In this work, we use the dependence of the relaxation time on the space between the echoes and the Larmour frequency (studied by Kleinberg and Horfield [4] and by Borgia et. all. [5]).

In the first section of our work we defines the classical Hausdorf momentum problem and we show when we can extract stable results from multiexponential decreasing curves.

In the second section, we present an application of our method to measure directly, in a heterogeneous media, the distribution function for local magnetic field gradient and distribution function of difference of magnetic susceptibility, both derived to the stable reconstruction of T_2 distribution. We measure the effect of:

a) diffusion in liquids mixture (acetone-water),
b) reduction of motion trembling and susceptibility variation (water saturated silica),
c) local magnetic field gradient, for biological solution of amino acids.

2 METHOD AND RESULTS

2.1 Theoretical background

We prove that the full reconstruction of the NMR spectrum from experimental data is mathematically unstable, but there is an infinity of physical quantities that may be computed in a stable manner. We construct the simplest and most stable "weighted T_2"=T_{2W} and other higher moments.

2.1.1 Statement and reformulation of the problem

The NMR echo signal $s(t)$ usually is assumed to have a multi-exponential normalized ($s(0)=1$) form:

$$s(t) = \sum_j n_j \exp(-t/T_{2j}), n_j \geq 0; \quad \sum_j n_j = 1; \quad T_{2j} \geq 0; \tag{1}$$

without any explicit bounds on N_{max}. We considered that the NMR signal $s(t)$ sampled at equidistant point (for simplicity) $t=0,\Delta\tau,2\Delta\tau,N\Delta\tau$. Denote $s_k=s(t_k)=s(\Delta\tau \cdot k); \ k=0...N$. Our problem is to investigate what kind of rigorous information can be extracted about n_j and T_{2j} from (1). In the forthcoming, we will prove that despite the problem is far to be as simple as the first sight, we can obtain some workable results. Let's purpose we have same eventually additional information: $0 \leq T_{2min} \leq T_{2j} \leq T_{2max} \leq \infty$ (the extreme situation $T_{2min}=0$ or $T_{2max}=\infty$ we consider included). The $s(t)$ from (1) may be rewritten as a Stiéltjes-Lebesque integral:

$$s(t) = \int_{T_{2min}}^{T_{2max}} \exp(-t/\tau)d\rho(\tau) \quad , \quad s_k = \int_{T_{2min}}^{T_{2max}} \exp(-k\Delta t/\tau)d\rho(\tau) \tag{2}$$

In fact (2) is the most general form, because it includes also the continuos components of the spectrum. This form includes (1) as a special particular case, when $\rho(\tau)$ is a piecewise constant, monotonic, function. Then, in order to reproduce (1) from (2), we choose the jumps points of $\rho(\tau)$ to coincide with the set T_{2j}, and $\lim_{\varepsilon \to +0} (\rho(T_{2j} + \varepsilon) - \rho(T_{2j} - \varepsilon)) = n_j$. Thus (2) is more general then (1) (it may contain also a continuous part, which may be considered to appear because of continuous external perturbation of some species of atoms). In the following we assume only $\rho(\tau)$ to be monotone, non-decreasing, function. The problem is to extract maximal useful information from s_k, $(k=0...N)$ about n_j and T_{2j}. We can reformulate mathematically in the following manner: given $\rho(\tau) \geq 0$, at finite number of points $s_k=s(t_k)=s(\Delta\tau \cdot k); \ k=0...N$, compute $d\rho(\tau)$ or find maximal information about $d\rho(\tau)$. By changing the variable $\Delta t/\tau$; $d\rho(\Delta t/\tau)=dv(u)$ we obtain:

$s_k = s(t_k) = \int_{v_{min}}^{v_{max}} \exp(-ku)dv(u)$. By further change of variable: $exp(-u)=v$, $dv(u)= d\sigma(v)$ our

problem become the problem P1: Let $s_k = \int_{v_{min}}^{v_{max}} v^k d\sigma(v)$; $k = \overline{0,N}$, $N < \infty$, given $d\sigma(v)>0$

we find maximal information about $d\sigma(v)$, where $0 \leq v_{min} < v_{max} < 1$. This is known in mathematical as the incomplete Markov momentum problem (see [6] and the exhaustive references therein), in contradistinction to ideally $N=\infty$ case, known as the Hausdorf momentum problem.

2.1.2 Negative results

The P1-problem has a great number of applications both in pure and applied mathematics and was treated almost completely. The P1-problem with physically correct data has an infinity of solutions, the jump points of $\sigma(v)$ (i.e. the values of T_{2j}) may be everywhere on the (v_{min}, v_{max}). This means that T_{2j} may be everywhere on (T_{2min}, T_{2max}). Thus, the P1-problem must be reformulated. We note that we supposed that s_k is known with absolute precision. In fact, s_k is known only approximately and the small errors give rise to supplementary uncontrolled oscillations in any, more or less rigorous, pragmatic attempts to reconstruct (n_j, T_{2j}) or the solution may do not exist at all, when the errors increase. The instability of the P1 problem is similar with a large class of so called "ill posed problems" in the Jacques Hadamard sense: extrapolation of analytic functions to the boundary of analytic domain, solution of Laplace equation with Cauchy data, etc.

2.1.3 Stable results

Considering the following problem, the problem P2, $s_k = \int_{v_{min}}^{v_{max}} v^k d\sigma(v); \quad k = \overline{0,N} \quad d\sigma(v) > 0,$

find the extremal values of the integral: $I(\Omega) = \int_{T_{2min}}^{T_{2max}} \Omega(\tau) d\rho(\tau)$. In [6] it is proved that the problem P2 has the stable solution (good dependence of errors) and moreover, the difference between maximal and minimal value tends to zero when $N \to \infty$, the accuracy at fixed N increases with the smoothness of $\Omega(\tau)$. Unfortunately, the direct use of mathematically complete, analytic, very elegant classical methods is obstructed by experimental errors which give rise to huge numerical instabilities and needs further studies. Therefore we restrict ourselves to a very special but workable case. For practical purposes we can compute easily, setting $\Omega(\tau) = \tau^n$, directly.

Let $s(t)$ be an interpolation of $s(\Delta \tau \cdot k)$, $s(t) = \int_{T_{2min}}^{T_{2max}} \exp(-t/\tau) d\rho(\tau)$. Then

$\int_0^\infty e^{-\mu} s(t) dt = \int_{T_{2min}}^{T_{2max}} \dfrac{1}{\lambda + \dfrac{1}{\tau}} d\rho(\tau)$. Differentiating n-times with λ and setting $\lambda = 0$

$\int_0^\infty t^n s(t) dt = n! \int_{T_{2min}}^{T_{2max}} \tau^{n+1} d\rho(\tau)$ and $\int_0^\infty Q_n(t) s(t) dt = n! \int_{T_{2min}}^{T_{2max}} \tau Pn(\tau) d\rho(\tau)$, where $P_n(\tau)$ is any

polynomial of degree n, $P_n(t) = \sum_{k=0}^n p_k t^k$ and $Q_n(t) = \sum_{k=0}^n \dfrac{p_k}{k!} t^k$. In the numerical

computation of $\int_0^\infty Q_n(t) s(t) dt$ despite $s(t) \xrightarrow{t \to \infty} 0$, for $n > 0$, a numerical instability $(P_n(t) \to \infty)$ occurs. Thus we used the most stable and simplest possible setting, $Q_n(t) = P_n(t) = 1, t, t^2, t^3$ in our work. Of course, there is an infinity of another stable quantities that can be extracted from NMR signal. According to the most researchers, the common relaxation curve (signal $s(t)$) to investigate T_2 in complex systems like gels, sol, emulsions, cellular tissue is multi-exponential, represented by (1) where n_i is population of compartments with T_{2i} relaxation time. The true number of these exponential is very difficult to predict experimentally because the level of noise is high and the relaxation phenomena are very complex. In this case, using (1) approximation, T_{2w} represents weighted mean for spin-spin relaxation time:

$$T_{2w} = \langle T \rangle = \int_0^\infty s(t) dt = \sum_i n_i T_{2i}, \quad \langle T^2 \rangle = \int_0^\infty t \cdot s(t) dt \qquad (3)$$

$$\langle T^3 \rangle = \int_0^\infty t^2 \cdot s(t)dt, \quad \langle T^4 \rangle = \int_0^\infty t^3 \cdot s(t)dt.$$

And we can construct the dispersion, asymmetry and kurtosis of T_2 distribution.

The $t = \infty$ limit is practically substituted with the time when $s(t)$ is less than the noise. The influence of noise in these parameters is not significant. The values T_{2W} have been determined by two methods, in test experiments. First, we integrated ordinarily in time domain, and second, we reconstructed the relaxation curve by usual multi-exponential fit [7] and calculated (3). We noted that these two methods give similar results.

2.2   Experimental background

For the heterogeneous sample, the internal magnetic field gradients have a large distribution of values and therefore the measured diffusivity in the sample does not correspond to the true diffusion coefficient. These two effects, the internal magnetic field gradients and the various values of the diffusivity of the wall surface lead to the difficulty of interpretation of the relaxation and diffusion model near the macromolecules.

In this work we have used CPMG pulse sequence in order to investigate diffusion effects in presence of spatially variation of main magnetic field on ox direction, g. In the conditions of free diffusion, the signal starts to decrease exponentially accordingly to the well-known expression [8]:

$$M(t) = M_0 \, e^{-\frac{t}{T_2}} \, e^{-\frac{1}{12}D\gamma^2 g_x^2 t_E^2 t} \tag{4}$$

, where t is k-echo time $(t=k \cdot t_E)$, D is the diffusion constant, g_x is the gradient of the magnetic field and T_2 is the intrinsic spin-spin relaxation time. We shall focus on the dependence of the decreasing signal as a function of the echo time t_E.

How can the above observation be generalised? In the heterogeneous systems, the magnetic field variations are even more complicated than those described by a constant gradient. In the absence of an utilisable theory of the inhomogeneous gradients, the practical conclusion of utilising some local gradient g, as an average of specific local gradients, is obvious. Therefore, the damping of the echo, due to the diffusion of the spins in the molecular fields' gradients can be modelled as a distribution of local gradient $f(g)$ of the sample [9]:

$$M(t) = M_0 \int f(g) \, e^{-\left(\frac{1}{T_2} + \frac{1}{12}D\gamma^2 g^2 t_E^2\right)t} \, dg \tag{5}$$

This expression has a major inconvenience: it is very complicated in order to estimate simultaneously the diffusion coefficient D, the distribution function $f(g)$ and the local gradient g. We can introduce the susceptibility difference $|\Delta\chi|$ in order to reduce the mathematical complexity and to obtain a better understanding.

For self-diffusion in heterogeneous system, the fluctuation of correlation time τ_c, is of order l_s^2/D, where l_s are the correlation length for field variation and D is the diffusion constant. The size of this fluctuation is of the order $\gamma |\Delta\chi| Bo$ where $|\Delta\chi|$ is the susceptibility variation at boundaries, with typically value in the range of 10^3-$10^4 s^{-1}$. The rapid magnetic field fluctuations (fast regime) result in homogeneous broadening (attenuation exponent inversely proportional to D) that is irreversible by spin echo method. Slow fluctuations can be refocused by a spin echo because result in the inhomogeneous broadening. The spin echo attenuation decrease with $exp(-2/3 \, \gamma^2 G^2 D\tau^3)$, where G is the mean of the local magnetic field gradient, similar to the usual spin echo attenuation when a uniform external magnetic field gradient is applied. Thus, the local susceptibility can be treated as an internal contribution to the applied gradient magnetic field, g_x, giving the total non-uniform magnetic field gradient as $g(r,t)=g_x(t)+G(r)$. This g is associated with $|\Delta\chi|$ through the expression $g \sim factor \cdot \left(\dfrac{\gamma}{D}\right)^{1/2} \left(|\Delta\chi| B_0\right)^{3/2}$ [10]. Therefore, when the factors are of order unity, we can write the expression (5) as follows:

$$M(t) = M_0 \int d|\Delta\chi| \, F(|\Delta\chi|) \, e^{-\left(\frac{1}{T_2}+\frac{1}{12}\gamma^3|\Delta\chi|^3 B_0^3 t_E^2\right)t} \tag{6}$$

We have used the equation (6) in order to extract the distribution $F(|\Delta\chi|)$ from CPMG decreasing curve.

2.3 NMR measurements

The time domain NMR measurements (TD-^{1}HNMR) were performed on a 25MHz ^{1}H-NMR AREMI 78 pulse spectrometer (manufactured by the Institute of Physics and Nuclear Engineering, Bucharest-Magurele, Romania) equipped with an electromagnet (110mm poles diameter). The overall stability of the magnetic field is 10^{-6}T/min and the magnetic field uniformity is 0.1G/cm in sample volume (1cm^3). For an accurate determination, we use a quadrature phase sensitive detector. The duration of the 90^0 pulse is less than 2µs. For diffusivity measurements we use the gradient of the main magnetic field oriented to *ox* direction $B(z)=B_0+g_x \cdot x$. The value of g is *1.8±0.2G/cm*. All measurements were carried out at *25±0.1^{0}C*. For CPMG sequence, we use various echo times (*0.8, 1.2, 1.6, 2, 2.4, 3.2, 4, 4.8, 6.4, 8, 11.2, 14, 18, 22.4, 44.8ms*) corresponding to the various experimental points (16000 at 300). The repetition delay (RD) was set to 15 sec and the enhancement was 6 (36 scans with S/N ratio~70dB). By quantitative analyses, we establish that all the protons in the sample (water protons, OH protons and aliphatic protons) contribute to the NMR signal [11]. The observed value of T_2 is characteristic to the overall protons in the sample [12]. The intrinsic spin-spin relaxation time was calculated after the extrapolation of the curve $T_2^{-1}(t_E)$ at very short t_E *(100µs)*.

For experimental measurements, we have used pure substances (Sigma-Chemie). First, we measured the relaxation curves for acetone and water (vol:vol) together, filled separately in capillary tubes and second, we measured the mixed solution of acetone with water (vol:vol) for all 15 values of t_E. The final form of convolution is obtained through averaging the all convolution reconstructed for each t_E.

When acetone and water are filled in capillary tubes we can decompose the relaxation curves in sum of two exponential, each exponential give us the specific contribution of diffusion to the relaxation process. As we expected, we obtained for pure acetone and for water one peak for the distribution function of local gradient, centred to the value of gradient g, assuming that diffusion coefficient for water is $2.3 \cdot 10^{-5} cm^2/s$ and for acetone $4.2 \cdot 10^{-5} cm^2/s$. We can extract intrinsic relaxation time and draw the convolution of local gradient (see Figure 1A). This experiment gives us information about how accurate is the computing method to this S/N ratio. It is easy to see that measuring and computing together the multiexponential decreasing signal the abundance of parameters who must be extracted very accurate (intrinsic relaxation time, distribution function) depend on diffusion coefficient and for this the computing error increases very much at long t_E. The same effect is present to the mixed solution of acetone and water when besides the contribution discussed above, appear the influence of diffusivity of water in relaxation curves of acetone (see Figure 1B). The modified distribution function for acetone at small value of local gradients is due to the fraction of spins with intrinsic relaxation time of acetone and diffusion coefficient of water. This modification of parameters move the distribution peak to the applied value of external gradient. This is also possible when in solution was formed the acetone hydrate.

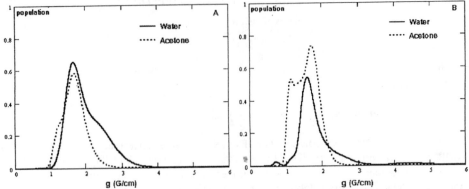

Figure 1   *The convolution of g -distribution for water (solid line) and acetone (dot line): A)with substances in capillary tubes are measured together (vol:vol) and B) for solution of acetone and water (vol:vol).*

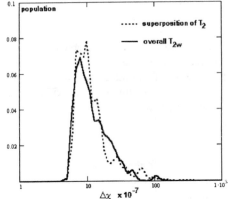

Figure 2   *The convolution of $|\Delta\chi|$ distribution of water-saturated silica powder obtained by two methods, superposition of multiexponential and evolution of overall parameter T_{2W}.*

In the second set of experiments, CPMG measurements were performed on water-saturated silica powder at different t_E. When we are dealing with the high complex system, with very strong influences between compartments of spins is very useful the stable parameter like T_{2w} (diffusion coefficients, local gradients and intrinsic relaxation time of spins are each characterised by distribution functions). All variation of all relaxation times is cumulated by T_{2w}. In water-saturated powder, the transverse relaxation rate is dominated by surface relaxation and relaxation caused by gradients only dominates for larger t_E. The surface relaxation can be described by a distribution of relaxation times T_2 . In this case, we can apply the equation (6), modified for distribution of relaxation time T_2. We obtain decomposition of CPMG curve in four exponential, with corresponding populations. In Figure 2 we show the superposition of $F(|\Delta\chi|)$ for four exponential and $F(|\Delta\chi|)$ extracted from the evolution of T_{2w}. These convolutions show us similar behaviour and we can say that the two different methods give similar results.

The most interesting result of our experiments is extraction of distribution function of local gradient for solution of enantiomers of amino acids (tryptophan, asparagine). Chiral molecules are complex enough molecules capable of supporting life and thus the role of chirality is very important in all the stages of life, from its appearance up to biochemical complex phenomena, such as metabolism. It is easy to understand that at the beginning of life it was necessary to produce a breaking of symmetry so that only one of the chiral forms of the same molecule should appear further on giving rise to homochirality [13]. Which are the elements that lead to such a strong chiral selection? The effect of the parity violation is very small, and yet it appears at the molecular level [14]. Since the real enantiomer of the L-amino acid is D-amino acid of the antimatter [15] we can say that L and D are not absolutely enantiomer molecules. Besides the dynamic effects given by the chemical exchange and the proton autodiffusion constant, a series of other properties connected to chirality are pointed out in Barra's [16] paper as shield constant σ and spin-spin coupling constant J. At the same time, the magnetic field, applied to a photo-induced biochemical reaction leads [17] to the conclusion that there is a connection between chirality–homochirality and the magnetic field.

In this particular case, the slight difference between intrinsic relaxation times of sorts of diffusive water near to macromolecule and possible difference in diffusion coefficients lead us to applied the overall method with T_{2W}. In Figure 3A,B we shows the specific behaviour of local gradients for D, L and racemic solution of tryptophan with concentration of *6.25mg/ml* and D,L solution of asparagine (*30mg/ml*). We use diffusion coefficient of water, presented above, for this calculus. The high value of local gradients results from calculus and this is representation of real local gradients corrected with the efficiency in relaxation suggested in [18].

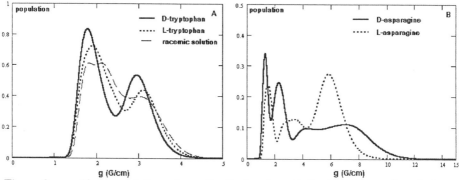

Figure 3   *The convolution of g -distribution for: A) D, L and racemic solution of tryptophan (pH 7.2) and B)D, L solution of asparagine.*

The slight difference between peaks (*g* mean) of enantiomers in solution is the order of *0.07G/cm* for tryptophan and even more for asparagine. This value give us an energy difference of $10^{-17}kT$ per molecule, comparative with PVED (parity violated energy difference). We note that similar behaviour we obtain with glucose, ribose, lactic acid [19]. The further investigations are centred to this mechanism that can be important in many biochemical processes like molecular recognition.

3   CONCLUSIONS

We proved that T_{2W} is an parameter less sensitive at noise level comparative to the non-linear regression of relaxation curves and also is very fast to compute. This parameter is a measure of the global proton relaxation in sample because the changes of quantity and quality of the relaxation compartments are reflected in the T_{2W} value.

We proved that when applied the algorithm of stable reconstruction of T_2 distribution on a liquid mixture of acetone and water (vol:vol), we find interaction between water and acetone. This interaction is represented by appearance of new peak in distribution function of acetone characterised by intrinsic relaxation time of acetone and diffusion coefficient of water.

In case of porous systems we can applied same algorithm of stable reconstruction and we prove that diffusive process affect the accuracy of this reconstruction but similar results we obtain with overall investigation with T_{2W}.

The most important results is difference in distribution function of local gradient between D and L enantiomers of amino acids. This difference in energy is linked by PEVD in magnetic field for explain the origin of homochirality.

References

1   Gy. Steinbrecher, R. Scorei, V.M. Cimpoiasu, I. Petrisor., *J. Magn. Reson.*, 2000, **146**, 321.

2   M.D. Hurlimann and D.D. Griffin, *J. Magn. Reson.*, 2000, **143**, 120.

3   J. Zhong, R.P. Kennan and J.C. Gore, *J. Magn. Reson.*, 1991, **95**, 267.

4   R.L. Kleinberg, M.A. Horsefield, *J. Magn. Reson.*, 1990, **88**, 9.

5   G.C. Borgia, R.J.S. Brown, P. Fantazzini, *Phys. Rev. E* , 1995, **51**, 2104.

6   M.G. Krein, A.A Nudemann, in: *Markov momentum problem and extremal questions*, Ed. Nauka, Moskow, 1973 (in Russian).

7   D. N. Rutledge, in: *Signal treatment and signal analysis in NMR*; D.N. Rutledge eds., Elsevier, Amsterdam, 1996, p. 191-217.

8   H.C. Torrey, *Phys. Rev.* , 1956, **104 (3)**, 563.

9   M.D. Hurlimann, *J. Magn. Reson.*, 1998, **131**, 232.

10 R.J.S. Brown, P. Fantazzini, *Phys. Rew. B*, 1993, **47**, 14823.

11 R. Scorei, V.M. Cimpoiasu, I. Petrisor, M. Iacob, I. Brad, I. Olteanu, R. Grosescu, V. Scorei, M. Mitrut and R. Cimpoiasu, *Rom. J. of Biophysics*, 1997, **4**, 327.

12 V.M. Cimpoiasu, Gy. Steinbrecher, I. Petrisor, R. Grosescu, R. Scorei, I. Brad, I.Olteanu, M. Iacob, Vilma Scorei, Mihaela Mitrut and Rodica Cimpoiasu, *Physics AUC*, 1997, **7**, 56.

13 W.A. Bonner, *Origins of Life and Evolution of the Biosphere*, 1995, **25**, 175.

14 W.A. Bonner, *Origins of Life and Evolution of the Biosphere*, 1996, **26**, 27.

15 A.J. MacDermott, G.E. Tranter and S.J. Trainor, *Chem. Phys. Lett.*, 1992, **194**, 152.

16 A.L. Barra, J.B. Robert and L. Wiesenfeld, *Phys. Lett. A.*, 1986, **115**, 443.

17 G.L.J.A. Rikken and E. Raupach, *Nature*, 1997, **390**, 493.

18 M.D. Hurlimann, *J. Magn. Reson.*, 2001, **148**, 367.

19 R. Scorei, V.M. Cimpoiasu, I.Petrisor and Gy. Steinbrecher, in: *Final program & Book of Abstacts* , P18, Symposium on Biological Chirality, 2000, Szeged, Hungary.

THE USE OF 2D LAPLACE INVERSION IN FOOD MATERIAL

S. Godefroy[1], L. K. Creamer[2], P. J. Watkinson[2] and P. T. Callaghan[1]

[1] MacDiarmid Institute for advanced materials and nanotechnology, School of Chemical and Physical Sciences, Victoria University of Wellington, Wellington, New Zealand
[2] Fonterra Research Centre (formerly known as the New Zealand Dairy Research Institute), Palmerston North, New Zealand

1 INTRODUCTION

In our explorations of the structural changes that occur within natural Mozzarella cheese during ripening, the changes to the water structure in different parts of the cheese has proven difficult to measure at the micron level. One possibility was to measure the sizes of the water 'compartments' by proton diffusion and relaxation using NMR techniques. It was anticipated that the protons in very small pools of water, for example, would show a lowered diffusion (because of restricted motion) and a normal relaxation.

Callaghan et al[1] have shown that at 30C the fat in Cheddar cheese was diffusion-restricted and was most probably associated with small droplets whereas the water diffusion was confined to surfaces within the protein matrix. MacGibbon et al (personal communication) has more recently shown that the size distribution of small aqueous droplets (ca. 5 micron) could be determined in butter or margarine using pulse gradient methods.

In the present study we were able to capitalise on novel computational methods that allow us to obtain, by a two-dimensional (2-D) Laplace Inversion[2], 2-D distribution maps of diffusion and spin relaxation (T_1 and T_2). NMR spin relaxation measurement is an efficient way of probing molecular interactions and the dynamics of molecular tumbling. Diffusion measurements by Pulsed Gradient Spin Echo (PGSE) technique allow one to explore the translational molecular motion[3]. Two dimensional spin-relaxation time and/or diffusion coefficient experiments (T_1/T_2 or $T_{1,2}/D$) can be used to correlate the different molecular dynamics and/or interactions by measuring the diffusion or relaxation parameters in a joint encoding. For example, the T_1 vs. T_2 correlations experiments might be used to elucidate the relative importance of different molecular tumbling rates. The importance of those correlations relies mostly on the determination of the ratio between T_1 and T_2. T_1 represents the time taken to establish thermal equilibrium and is dominated by high frequency tumbling while T_2 represents the time taken to dephase and is dominated by low frequency tumbling. The D vs. T_2 correlations allow molecular motion to be distinguished from molecular interactions experienced by the system of spins. Because of the heterogeneity of the materials under study, a wide range of relaxation and diffusion is observed. The two-dimensional data are 2-D multi-exponential decay plots and were therefore analysed by using a 2-D Laplace Inversion.

2 EXPERIMENTS

2.1 Samples

The purpose of the experiments performed in this study was to investigate the water and fat in cheese and to determine whether T_1/T_2 or T_2/D plots could act as a marker for cheese properties. In particular, we have sought to investigate these plots as a function of cheese temperature and age. A dry-salted Mozzarella-style (pasta filata or hot-stretched) and a dry-salted Gouda-style cheese were manufactured specifically for this project. One of the two 20 kg blocks of cheese was cut into 16 nominally 1 kg blocks and these were individually vacuum-packed and stored at 5C until assayed or examined by NMR spectroscopy.

Prior to each NMR measurement a vacuum-packed block was opened and a small sample taken from the middle portion (and in the case of Mozzarella-style cheese, parallel to the direction of stretching). The sample used for NMR measurement was 20 mm long and fitted tightly into a 10 mm diameter glass open-ended NMR tube. Tight polytetrafluoroethylene plugs were then positioned on each side of the cheese plugs to reduce cheese moisture loss by evaporation to the air. Evaporation of water was negligible with this sample packing, as shown by the same sample mass before and after typical NMR experiments of several hours duration. NMR data were collected from the two cheeses at 5 or 40 C at ages between 1 to 74 days. The composition of each cheese 1 day after manufacture (Table 1) was within the expected range.

Table 1 *Composition of the cheeses*

	Fat (%)	Moisture (%)	pH (-)	Salt (%)
Mozzarella-style	22.0	46.6	5.49	1.34
Gouda-style	31.5	38.3	5.33	1.63

2.2 NMR techniques

The NMR experiments were performed on an AVANCE and AMX 300 MHz Bruker spectrometer. In order to get information on the correlations between the diffusion coefficients and the transverse relation times and between the two relaxation times T_1 and T_2, we performed 2D MNR experiments by using the NMR pulse sequence as shown in the figures 1 and 2 below. The diffusion / relaxation correlation sequence consists of a combination of a Carr-Purcell-Meiboom-Gill (CPMG) sequence and a Pulsed-Gradient Spin-Echo (PGSE) sequence. As the spin-spin relaxation times (T_2) are more than one order of magnitude shorter than the spin-lattice relaxation times (T_1), we applied the PGSE sequence with stimulated echoes. The inter-pulse gradient delay Δ was chosen such that the water diffusion coefficient was independent of Δ, indicated no boundary effects on restricted diffusion[3]. The strength of the gradients applied in the 2D experiments was 0.5 T/m.

The water and oil signals were separated by means of their chemical shifts to provide independent data sets for each. The 2D data were analyzed using a two-dimensional Laplace Inversion[2], as described in the next section. The experiments were performed at 5 and 40°C.

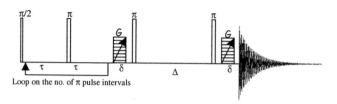

Figure 1 *D/T₂ correlation pulse sequence, as a combination of CPMG sequence and Pulsed Gradient Spin Echo (PGSE) sequence with stimulated echoes. The parameters of each dimension, namely the CPMG duration and the strength of the gradients G (or their duration δ), are varied independently.*

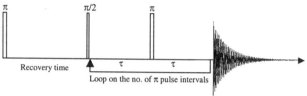

Figure 2 *T₁/T₂ correlation pulse sequence, as a combination of Inversion-Recovery and CPMG sequence. The recovery time and the CPMG duration are varied independently.*

2.3 Data processing

Due to the heterogeneity of the samples studied, we expect to acquire multi-exponential signals, both for the relaxation and for the diffusion dimension. Therefore, we developed a 2-D Laplace Inversion, using Matlab, based on the 1-D Laplace inversion[4,5], according to the algorithm of Song et al[2]. For 1D relaxation experiments, the signal $S(t_i)$ can by expressed by:

$$S(t_i) = y_i = \sum_k X(T_k) \exp\left(-t_i / T_{1,2k}\right) + \varepsilon_i \tag{1}$$

$T_{1,2k}$ and $X(T_{1,2k})$ correspond respectively to the relaxation times of the system and their distribution. ε_i is the noise. The equation (1) can be written on a matrix form, such as:

$$Y = KX + E \tag{2}$$

where K is the known matrix of the $exp(-t_i/T_k)$ (with t_i the time and T_k the relaxation times, fixed for the processing). To determine the distribution X, the general procedure consists in a non-negative least square fit[6], with an additional smoothing term, where the smoothing parameter $\alpha_,$, gives the weight of this additional term. We want to minimize χ^2 such as:

$$\chi^2 = \left\| KX + E \right\|^2 + \alpha^2 \left\| X \right\| \tag{3}$$

For the two dimensional diffusion (D) vs. relaxation (T_2) and relaxation (T_1) vs. relaxation (T_2) experiments as described on the figures 1 and 2, the acquisition signals are given by:

$$s(t_{i1},q_{i2}) = Y_{i1,i2} = \sum \exp(-t_{i1}/T_{2k1}).\exp(-q_{i2}^2 \Delta D_{k2})X(T_{2k1},D_{k2}) + E(t_{i1},q_{i2})$$
$$s(t_{i1},t_{i2}) = Y_{i1,i2} = \sum \exp(-t_{i1}/T_{2k1}).\exp(-t_{i2}/T_{1k2})X(T_{2k1},T_{1k2}) + E(t_{i1},t_{i2})$$

(4)

In the matrix form, both signals will be written as:

$$Y = K_1 X K_2' + E$$

(5)

where K_1 and K_2 are, as previously, the known matrices of $exp(-t_{1i}/T_{2k1})$ and $exp(-q_{i2}\Delta D_{k2})$ (or $exp(-t_{i2}/T_{2k2})$), respectively. To get the 2D Laplace Inversion of the 2D signal Y, the idea is to transform the equation (5) into a 1D equation (2), by transforming the matrix Y into a vector. However, this 2D→1D transformation leads to huge matrices, thus to very long computation times. Therefore, we first apply a singular value decomposition[2] (SVD) to the matrices K_1, K_2 and Y, to get the squared, reduced matrices $\tilde{K}_1$, $\tilde{K}_2$ and $\tilde{Y}$. The equation (5) can now be applied with the reduced matrices. We then apply the 2D→1D transformation on the reduced data, and process the data according to the 1D Laplace Inversion presented in the equations (2) and (3):

$$\tilde{y} = \tilde{K}x + e, \quad \tilde{y} = [\tilde{Y}]_D, \quad x = [X]_D, \quad \tilde{K} = \tilde{K}_1 \otimes \tilde{K}_2$$

(6)

3 RESULTS AND DISCUSSION

3.1 Mozzarella

Below are presented the water (on the left) and oil (on the right) two dimensional distribution maps of diffusion *vs.* relaxation at 5°C for the aging days 3 and 73 (figure 3), spin-lattice relaxation *vs.* spin-spin relaxation at 5°C for the aging age 25 (figure 4) and of diffusion *vs.* relaxation at 40°C at the aging days 18 and 73 (figure 5). The D *vs.* T_2 correlation map of water at 5°C and day 3 shows only one peak centred at $D = 1.8\ 10^{-10}$ m^2/s and $T_2 = 10$ ms. Neither diffusion nor relaxation times vary significantly with aging although the relaxation time increases slightly at day 73 to $T_2 = 15$ ms. For oil, we get two different peaks with T_2 value between 10 and 40 ms. Those peaks correspond mainly to solid fat, so that very little signal attenuation is observed in the diffusion measurement, with the gradient strength used. The diffusion coefficients of those peaks are below 10^{-12} m^2/s. As the cheese gets older, a third oil peak appears at much larger diffusion coefficients ($D = 1.7\ 10^{-10}$ m^2/s) but at similar relaxation time ($T_2 = 12$ ms). This peak provides evidence with aging of a small population of liquid fat molecules. The T_1 *vs.* T_2 distribution maps, showing spin-lattice relaxation times T_1 longer than the spin-spin relaxation times T_2, are typical for those soft media, with $T_1 = 470$ ms and $T_2 = 10$ ms for water and for the main peak for oil, $T_1 = 480$ ms and $T_2 = 15$ ms. We note that the water T_1 relaxation times slightly decrease with aging, whereas the T_2 values slightly increase, as was of course observed from the D *vs.* T_2 maps.

The measurements performed as a function of aging at 40°C seem to give more information. Indeed, for water, the population of spins with a low diffusion coefficient (D

= 2.2 10^{-11} m^2/s) and a high relaxation time (T_2 = 31 ms) seems to shift during aging to one with a higher D *(10^{-9} m^2/s) and* a much shorter T_2 (1.5 ms). This might be explained by a more efficient change of environment of the water molecules with the temperature when the cheese ages. Note that the diffusion coefficient of water for the young cheese (below 20 days) is smaller at 40°C than at 5°C. This unexpected result might be explained by a phase transition. For oil, the diffusion coefficients at 40°C are larger than at 5°C.

Previous spin-spin relaxation experiments from Kuo et al.[7] as a function of aging from day 1 to 10 carried out at 5°C have shown two relaxation times with the longer T_2 component dominating at first but the shorter T_2 component dominating with aging. Those results have shown that there is a redistribution of water with aging from a more- to less-mobile fraction.

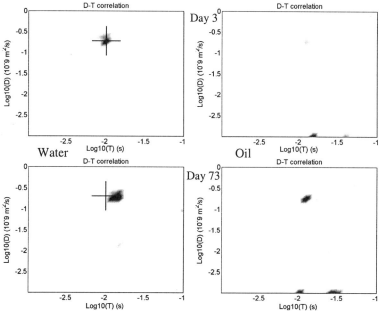

Figure 3 *D/T_2 correlation distribution maps for the Mozzarella samples, measured at 5 °C for the aging days 3 and 73, for water (left) and oil (right).*

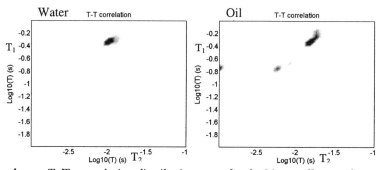

Figure 4 *T_1/T_2 correlation distribution maps for the Mozzarella samples, measured at 5 °C for the aging Day 25, for water (left) and oil (right).*

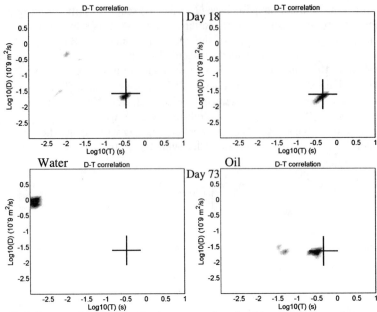

Figure 5 *D/T₂ correlation distribution maps for the Mozzarella samples, measured at 40 °C for the aging days 18 and 73, for water (left) and oil (right).*

3.2 Gouda

Below are displayed the water (left) and oil (right) 2-D distribution maps of diffusion *vs.* relaxation at 5°C for the aging days 3 and 73 (figure 6) and at 40°C at the aging days 10 and 73 (figure 7). The *D vs. T₂* correlation map of water at 5°C and day 3 shows only one peak centred at $D = 1.3 \ 10^{-10}$ m²/s and $T_2 = 7$ ms. As for the Mozzarella cheese, we observe little change for the main peak except that T_2 is increasing slightly from the third day of aging to reach the value of 10 ms at the end. For oil we observe very small diffusion coefficient, corresponding to the solid fat population that would require stronger gradients to be measured. This behaviour is identical to that seen for Mozzarella. The measurements of *D vs. T₂* performed as a function of aging at 40°C seem to give more information. For water, we observe once again an exchange from the population of molecules at low *D* and high T_2 to two other populations at low *D* and lower T_2 and higher *D* and lower T_2.

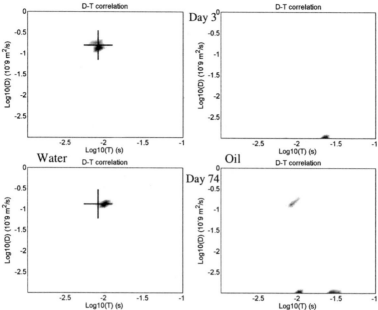

Figure 6 *D/T₂ correlation distribution maps for the Gouda samples, measured at 5 °C for the aging days 3 and 74, for water (left) and oil (right).*

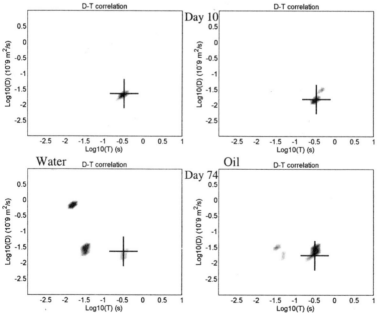

Figure 7 *D/T₂ correlation distribution maps for the Gouda samples, measured at 40 °C at the days 10 and 74, for water (left) and oil (right).*

4 CONCLUSION

The T_1 vs. T_2 and D vs. T_2 plots shown here indicate that correlated relaxation and diffusion experiments analysed by 2-D Laplace Inversion can reveal new features in oil and water dynamics for food systems. In particular, at 40°C we observed changes in those correlations as cheeses age. By contrast, at 5°C the changes of T_2 but not D for water as a cheese ages, shows that the simpler 1-D analysis is adequate in this case.

The full explanation for the differences between the patterns at 5 and 40°C and the impact of aging is not obvious and requires further investigation. However, we believe that these differences may arise from a change in phase of the water possibly associated with protein bonding phenomena.

The computational methods developed for the current study of water and oil in cheese should be transferable to other food systems. They have opened up possibilities of multi-dimensional display and analysis of multi-phase food products.

Acknowledgements
The authors would like to thank the Fonterra Research Centre for major funding of this project and for providing the cheese samples.

References
1.   P. T. Callaghan, K. W. Jolley and R. S. Humphrey, diffusion of fat and water as studied by pulsed filed gradient nuclear magnetic resonance, J. Colloi. Inter. Sc., 1982, **93**, 521.
2.   Y.-Q.-Song, L. Venkataramanan, M. D. Hurlimann, M. Flaum, P. Frulla and C. Straley, T1-T2 correlation spectra obtained using a fast two-dimensional Laplace Inversion, Journal of Magnetic Resonance, 2002, **154**, 261.
3.   P. T. Callaghan, Principles of nuclear magnetic resonance microscopy, Clarendon Press, Oxford, 1991
4.   S.W. Provencher: CONTIN: A general purpose constrained regularization program for inverting noisy linear algebraic and integral equations. *Comput. Phys. Commun.* 1982, **27**, 229.
5.   G. C. Borgia, R. J. S. Brown and P. Fantazzini, Uniform-Penalty Inversion of Multiexponential Decay Data, Journal of Magnetic Resonance, 1998, **132**, 65.
6.   C. L. Lawson and R. J. Hanson, solving least squares problems, Englewood Cliffs, N.J., Prentice-Hall, 1974.
7.   M.-I. Kuo, S. Gunasekaran, M. Johnson and C. Chen, Nuclear Magnetic Resonance Study of water mobility in Pasta Filata and non-Pasta Filata Mozzarella, J. Dairy Sci. 2001, **84**, 1950.

SPIN-SPIN AND SPIN-LATTICE RELAXOMETRY IN SUNFLOWER SEEDS

G. Albert[1], D.J. Pusiol[2] and M.J. Zuriaga[2]

[1]FOINTEC SRL, A.M. Bas 236, P.A. Dpto 4B, X5000KLF Córdoba, Argentina.
[2]Facultad de Matemática, Astronomía y Física, Universidad Nacional de Córdoba, Ciudad Universitaria, X5004DVA Córdoba, Argentina. E.mail: pusiol@famaf.unc.edu.ar

1 INTRODUCTION

Low resolution Magnetic Resonance techniques has been used for more than thirty years to analyze the moisture:oil:solid matter ratios in oilseeds. The relative contents were, and still are obtained from the evaluation of the intensity ratios of fast (solids) medium (oil) and slow (water) decays of the Free Induction Decay (FID) signal. This procedure require very good magnetic field homogeneity, in addition with a very well controlled magnetic field stability during the measure process. Permanent magnets (PMs) -which are commonly used in low resolution spectrometers- are technologically easy to implement than electromagnets. PMs do not need sophisticated and expensive field control electronics; but thermal shifting of the magnet dimensions produced by changes in the room temperature as well as by the rf pulses, can not be avoided. Thermalization of the hundred of Kg of PMs needs to be maintained during the spectrometer working time.

Recently Carr-Purcell-Meiboom-Gill (CPMG) sequence[1,2] became a standard procedure for measuring transverse relaxation times. It is well known that the Meiboom-Gill modification[2] compensate for small inhomogeneities in the static field, B_0, and in the rf field strength, B_1. This makes possible to refocus the transverse magnetization many times, limited only by the intrinsic relaxation time of the sample. Moreover, CPMG can still be used to measure the distribution of relaxation times[3,4] in inhomogeneous fields as well as in the off-resonance condition. It was recently shown that, by implementation of the CPMG sequence with composite pulses, the spin-spin relaxometry profile can also be acquired by spectrometers with strong B_0, B_1 inhomogeneities together with closely untuned rf irradiation[5]. Spectrometers with light PMs and working at room temperature can now be used to collect the decay of well focalized echoes.

In this work, the application of the CPMG pulse sequence to well known sunflower seeds with the aim to resolve the T_2-dispersion profiles within the oil magnetization decay signal is presented. On the other hand the relaxometry of the longitudinal magnetization is analyzed. The aim here is the study of the fatty acid molecular dynamics inside the seed structure. Spin-lattice relaxation time studies has been regularly used, for more than 20 years, to analyze the reorientations of molecules in complex molecular systems[6]. The underlying principle is that the relaxation rates reflect the intra and the intermolecular interactions of nuclear spins with local magnetic fields which are, in turn, modulated by molecular motions[7]. Different types of process, such as rotational jumps or reorientation mediated by spatial translations can be distinguished by analyzing the Fourier spectra of the time dependent interactions[7]. In order to do this, it is important that the NMR experiment should provide a sufficiently broad frequency window. In practice, conventional NMR spectrometers are restricted to a relatively small frequency range (in the MHz region). This narrow window is hardly adequate for a detailed evaluation of the motional spectrum. The situation is strongly improved by conducting experiments at different magnetic fields. This is what is call spin-lattice relaxometry[8]. Since the macroscopic magnetization must be preserved in magnitude while switching from high to low field and *vice versa*, and since T_1 values in complex fluids are often rather short at low fields (on the order of milliseconds), applications of FC-NMR techniques to those systems generally requires special instrumentation.

2. EXPERIMENTAL

2.1 Samples:

Sunflower seed samples were provided by the Argentine National Institute for Agricultural Research (INTA). Calibrations of water, proteins and fatty acids contents were carried out following the conventional analytical chemistry methods.

2.2 T_1 and T_2 techniques and set up

T_2-profiles were measured using the conventional CPMG sequence, acquiring 8,192 spin-echoes in a commercial Fointec cr20: 20 MHz permanent magnet NMR spectrometer. The echo amplitude data were recorded at room temperature taking care that radio frequency do not shift the sample temperature more than few tenth of degrees. Single shot T_2-profiles with S/N ratio of about 50:1 were averaged 100 times.

Data were analyzed by means of a multiexponential algorithm (Multiexp), which fits the 8k data set of the echo-decay with 100 exponential functions.

Our FFC instrument is home made. The magnetic field strength is cycled by electronically switching the current through a special magnet [9]. In order to achieve a fast field cycle, the electronic circuitry must be a compromise between having a high voltage in the power supply and magnets with a small inductance and resistance [10]. With the development of high voltage MOSFET's switching and with the aid of an energy storage device [9] the 0.42 T high fields is switched on and off in 700 μs and 1ms respectively

[11]. The low inductance and resistance coil is an air cored homogenized Kelvin-type magnet. It features a variable conductor cross-section calculated by a Lagrange minimization formalism [12] and machined by purpose-built computer controlled equipment [11, 12]. The H_R relaxation field strenght and stability have been calibrated and monitored by applying a second radio frequency irradiation during the relaxation period in a water sample and recording the resultant signal during the detection period [10]. FC-NMR principles and techniques are discussed in more detail in refs. [8] and [13]. $T_1(\nu_L)$ errors are estimated within 5 %.

3. RESULTS AND DISCUSSION

T_2-profiles present similar behaviors in the whole samples. Five peaks are observed. Fig 1 represent a typical profile collected from sample #11. The small T_2 is found at about 1 ms; a second one is found at 20 ms. Both are differentiate of the high T_2 triplet because of their small amplitudes.

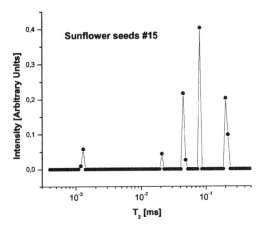

Figure 1: *The T_2 dispersion profile measured in sample 11. Five T_2 peaks are detected.*

The water peak is univocally assigned to the 20 ms decay, because it diminishes dramatically when the seeds are dried in the furnace. To assign the T_2 corresponding to the oil molecules, the profile of pure sunflower oil are measured. In order to compare, we depict the profiles of two samples in Fig. 2: *a)* a solution of water and cooper sulphate and *b)* 0.35 gr of pure sunflower oil. In the first case a single exponential decay is observed. From the second profile we can see that the transversal magnetization defocalizes following a triple exponential decay, just in the range of the high T_2 triplet measured in the seeds (Fig. 1). T_2 are respectively 44.2 ms, 86.4 ms and 219.2 ms.

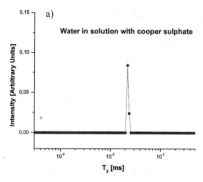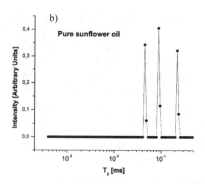

Figure 2 *Single and triple exponential decays are measured in the CPMG echo sequence respectively in doped water and in pure oil.*

Table I summarize the total data fitting of the five samples set. Note that T_2 values of the pure oil agree with the corresponding ones when the oil is located inside the seed.

#	% Oil	% Prot	% Moist	T_{2-1}	A_1	T_{2-2}	A_2	T_{2-3}	A_3	T_{2-4}	A_4	T_{2-5}	A_5
11	44.9	15.9	6.2	1.4	0.060	19.2	0.026	44.2	0.283	94.4	0.263	201.4	0.179
12	49.8	17.5	7.1	1.6	0.056	19.0	0.042	44.2	0.274	86.8	0.293	201.4	0.221
13	52.8	14.7	6.5	1.6	0.053	18.4	0.032	44.2	0.223	86.8	0.269	200.7	0.220
14	56.4	9.3	7	1.2	0.045	22.4	0.035	37.4	0.131	74.4	0.344	202.0	0.232
15	60.4	11.2	7.3	1.2	0.056	24.2	0.052	42.1	0.202	79.8	0.324	201.7	0.251

Table I *Fitting parameters obtained with the Multiexp program. The five peaks are represented by 1,...5. T_2's are expressed in ms.*

The same measurements were done in the second set of samples (# 18-20), which were previously calibrated by its respective fatty acid contents (See Table II). In Fig 3 we

Sample #	% Moist.	% Prot.	% Oil	%	Fatty	Acids		NMR $T_2 \sim 40$ ms	NMR $T_2 \sim 85$ ms
				16:0	18:0	18:1	18:2		
18	5.3	-	50.3	3.0	1.5	83.8	11.7	0.0349	0.011
19	6.7	-	45.4	5.3	1.9	26.3	66.5	0.0170	0.032
20	6.2	15.9	44.9	5.5	2.1	27.4	65.0	0.0200	0.029

Table II *Calibration parameters of samples 18-20 together with the amplitudes of the T_2 decays at about 40 ms and 85 ms, respectively (last two column right).*

show the correlation between the oleic (18:1) and linoleic (18:2) acids relative contents and the amplitude of the two T_2's here assigned: ~ 40ms to the first and ~ 85ms to the second, respectively. The correlation between the experimental points is noticeable.

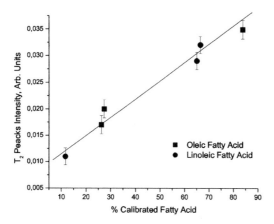

Figure 3 *Plot of the last two columns of Table IV vs. the oleic and linoleic fatty acid contents, obtained by physicochemical methods.*

Fast field Cycling measurements of T_1 carried out in sample # 20 is shown in Fig. 4. Two exponential decays fit the evolution of the longitudinal magnetization. More than 70 % of the signal in the FC apparatus comes from the unsaturated fatty acids; we assign these two decays to them.

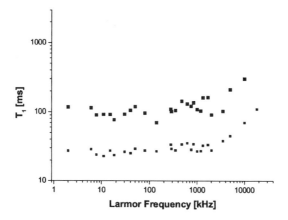

Figure 4 *$T_1(\nu_L)$-profiles of the two relaxation times measured in sunflower seeds. Both behaviour are similar to those observed in T_1 relaxometry of liquids.*

From the Larmor frequency dependence we note that the plateau in the $T_1(\nu_L)$ behaviours means that the dynamical state of fatty acid molecules is compatible with that in the liquid state. Therefore, from the point of view of the NMR, we can conclude that the oil into the seeds behaves like a liquid.

4. CONCLUSIONS

From this study it can be concluded that the CPMG pulse sequence allows the estimation of the relative contents of oleic and linoleic fatty acids by measuring the respective amplitudes of the echo decays with T_2 in the ranges of 40 and 85 ms. On the other hand we also conclude that the dynamical state of oil in the sunflower seeds can be assumed liquid-like. This feature was proved by the spin-lattice relaxometric study as well as by the comparison of the T_2-profiles of the seeds and pure oil.

Acknowledgments

Argentine National and Córdoba Provincial Agencies for Science and Technology CAC and ANPCYT are greatly acknowledged for financing partially the Project. DJP and MJZ hold a fellowship of the National Research Council, CONICET.

References

1. Carr H.R. and Purcell E.M., Phys. Rev., **94**, 630 (1954).
2. Meiboom S. and Gill D., Rev. Sci. Instrum., **29**, 688 (1958).
3. Hürlimann M.D. and Griffin D.D., J. Magn. Reson., **143**, 120 (2000).
4. Simbrunner J. and Stollberger R., J. Magn. Reson., **B109**, 301 (1995).
5. Brain A.D. and Randall E.W., J. Magn. Reson., **A123**, 49 (1996).
6. Emsley, J.W., *Nuclear Magnetic Resonance of Liquid Crystals*, Reidel (1985).
7. Abragam A., *Principles of Nuclear Magnetism*, Oxford Univ. Press, New York (1961).
8. Kimmich, R., *NMR: Tomography, Diffusometry and Relaxometry*, Springer Verlag (1997).
9. Rommel, E., Mischker, K., Osswald, G. Schweikert, K-H., and Noack, F., J. Magn. Reson., **70**, 219 (1986).
10. Anoardo, E. and Pusiol, D.J., Phys. Rev. Letters, **20**, 3983 (1996).
11. Anoardo, E., Thesis, Universidad Nacional de Córdoba (1996).
12. E. Anoardo and D.J. Pusiol, Phys. Rev. B, **56**, 2348 (1997).
13. D.J. Pusiol and E. Anoardo, Braz. J. Phys., **28**, 274 (1998).

Food: The Human Aspect

QUANTIFICATION OF LIPOPROTEIN SUBFRACTIONS USING ^{1}H-NMR AND CHEMOMETRICS

M. Dyrby [a], M. Petersen [b], S.B. Engelsen [a], L. Nørgaard [a] and U.G. Sidelmann [c]

[a] *The Royal Veterinary and Agricultural University, Centre for Advanced Food Studies, Department of Dairy and Food Science, Food Technology, Rolighedsvej 30, DK-1958 Frederiksberg C, Denmark*
[b] *The Royal Veterinary and Agricultural University, The Research Department of Human Nutrition, Rolighedsvej 30, DK-1958 Frederiksberg C, Denmark*
[c] *Novo Nordisk A/S, Applied Trinomics, Novo Nordisk Park, DK-2760 Måløv, Denmark*

1 INTRODUCTION

People suffering from Type II diabetes as well as dyslipidemia in general have an increased risk of coronary heart disease (CHD). The individual risk of CHD has shown to be related to the distribution of cholesterol and triglyceride in different types of lipoproteins.[1,2] It is therefore of utmost importance to be able to measure and monitor the lipoprotein profile in medical diagnostics as well as in relation to human nutrition studies and in testing of new drugs.

Lipoproteins are micellular triglyceride and cholesterol transport vehicles of blood and can be divided into subgroups on the basis of their density and size. The main fractions are Very-Low-Density-Lipoproteins (VLDL), Intermediate-Density-Lipoproteins (IDL), Low-Density-Lipoproteins (LDL) and High-Density-Lipoproteins (HDL). Figure 1 shows a schematic representation of a lipoprotein.[3]

Ultracentrifugation is the established standard reference method for the separation of lipoproteins. The partly empirical definitions of the main fractions are based on their density, VLDL having a density between 0.940 and 1.006 g/ml, IDL between 1.006 and 1.019 g/ml, LDL between 1.019 and 1.063 g/ml, and HDL between 1.063 and 1.21 g/ml.

Gel electrophoresis is a widely used alternative to ultracentrifugation for the separation and quantification of lipoproteins. The separation is based on a different principle, namely on the size and charge of lipopoteins, but apparently shows good correlation with ultracentrifugation.[4]

Lipoproteins in blood can also be measured using high-resolution NMR spectroscopy, but due to their similar composition, the subgroups have very similar – but not identical – signals. Furthermore, the signals are minimally shifted in frequency due to the different densities of the lipoproteins, giving rise to different local magnetic fields. The result is heavily overlapping peaks, and quantification of the different subgroups from the NMR spectra can only be done using multivariate data analysis (chemometrics).

For quantification, the most commonly used chemometric method is Partial Least Squares regression (PLS). This method is a very robust and powerful algorithm designed to build quantitative prediction models calibrated to samples with known composition.[5]

Several studies have attempted to quantify the main fractions using NMR with good correlations to standard methods,[6-10] while very few attemps have been made to quantify subfractions.[11] In addition, most studies have been performed using standard curve resolution and integration, while only few used the model-free chemometric approach.[8-10]

In the study presented here, 600 MHz ^{1}H-NMR was applied to plasma samples to evaluate the potential for quantification of lipoprotein subfractions using PLS. Using ultracentrifugation and gel electrophoresis as reference analysis methods, PLS prediction models based on both cholesterol and triglyceride in main fractions as well as several subfractions of lipoproteins were developed.

PL: phospholipid

FC: free cholesterol

CE: cholesterol ester

TG: triglyceride

Figure 1   *Schematic representation of a lipoprotein.[3]*
Diameters range from 5 to 80 nm.

2 EXPERIMENTAL

2.1 Plasma samples

Human volunteers with a broad range in lipid levels in terms of total triglycerides were chosen in order to span a resonable range of lipoprotein subfractions. Both males and females were included, and the subjects included obese, obese after five weeks of diet as well as lean controls.

Blood was drawn from 72 fasting subjects into EDTA tubes and centrifuged to separate the blood cells from plasma. The plasma was transferred to new test tubes and stored at − 80 °C until further analysis. All analyses were commenced no later than 24 hours after thawing. The samples were subjected to separation using ultracentrifugation, NMR analysis as well as gel electrophoresis.

2.2 Ultracentrifugation (UC)

The lipoproteins were subfractionated by serial ultracentrifugation as described in Baumstark *et al.* In short, the main fractions, VLDL, IDL, LDL and HDL, were separated from plasma one at a time, and LDL was subsequently separated into six subfractions (LDL 1 to LDL 6) and HDL into three subfractions (HDL 1 to HDL 3).[12] Furthermore, VLDL was separated into two subfractions (VLDL 1 and VLDL 2) as described in Caslake and Packard.[13] Part of the main fractions were set aside before subfractionation for further analysis, yielding a total of 15 fractions.

In the original plasma as well as in each main and subfraction, total cholesterol and triglyceride were determined by standard enzymatic colorimetric methods and apolipoprotein apoB100 by a standard imunoassay method using the Cobas Mira Plus instrument (Roche, Basel, Switzerland).

2.3   Gel Electrophoresis (GE)

Lipoprotein profiles were also determined using continuous polyacrylamide gel electrophoresis (Lipoprint LDL System, Quantimetrix, CA). In short, 25 μl EDTA plasma was added to each glass tube with 3 % continuous polyacrylamide precast gel. Loading gel containing Sudan Black (thus colouring for cholesterol) was added and mixed with the plasma, polymerized by fluorescent light and electrophoresed at a constant current of 3 mA for 60-70 minutes.

Gels in glass tubes were scanned directly at a wavelength of 610 nm. Different subfractions were identified by their migration distance. Each lipoprotein subfraction has a specific electrophoretic mobility (Rf) relative to the HDL fraction using the VLDL fraction at the origin of the separating gel and the HDL fraction at the end of the gel. The relative area of each lipoprotein subfraction was determined from the densiometric scan. This method quantifies cholesterol in VLDL, IDL 1 to IDL 3, LDL 1 to LDL 6 and HDL 2 to HDL 3, yielding a total of 12 fractions.

2.4 ^{1}H NMR

The samples were measured in 5 mm tubes containing 200 μl plasma, 600 μl 0.9 % NaCl in H_2O and 100 μl 27.5 mM formic acid in D_2O. Formic acid served as chemical shift reference, while D_2O provided deuterium field/frequency lock.

Proton spectra were measured at 30°C on a Bruker AV600 operating at 14.1 T. An 1D diffusion edited NMR experiment was performed using a diffusion time of 0.1 s in order to suppress signals from small molecules. After multiple pulse water presaturation a spectral window of 12376 Hz was accumulated in an acquisition time of 1.32 s. The relaxation delay was 2.0 s, the FID was collected into 65K data points and 128 scans were acquired on each sample.

Following acqustion, the FID's were Fourier transformed applying neither zerofilling nor window function. The spectra were manually phase-corrected, referenced to formic acid at 8.49 ppm (only partially suppressed by the diffusion experiment) and corrected for baseline offset errors using the flat regions at both ends of the spectrum. The spectral area chosen for multivariate analysis was 5.5 to 0.5 ppm, excluding the residual water region (4.5 – 5.1 ppm) and the region containing peaks from EDTA (3.8 – 2.6 ppm).

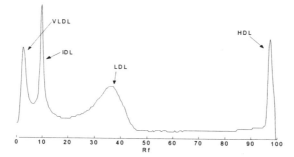

Figure 2   *Electropherogram of a plasma sample*

3 RESULTS

A representative electropherogram and a 1D diffusion-edited ^{1}H NMR spectrum of a plasma sample are shown in Figures 2 and 3, respectively. The NMR spectrum is shown before exclusion of water and EDTA spectral regions.

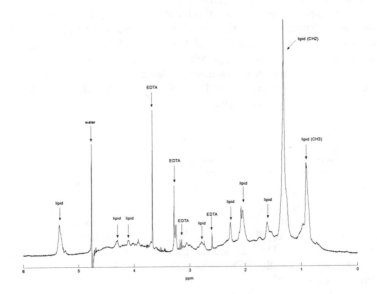

Figure 3   *^{1}H-NMR spectrum of plasma sample (water, EDTA and lipid peaks are marked)*

3.1 Lipoprotein subfraction quantification using UC as reference method

For each lipoprotein fraction a Partial Least Squares (PLS) regression model was developed between the NMR spectra and concentrations of cholesterol and triglyceride in the subfractions separated by UC.

For each model, outliers were removed and the optimal spectral range was evaluated. In order to determine the correct number of PLS components, cross validation was applied using ten segments.

Results of predictions using UC as quantitative reference method to the NMR data are shown in Table I. Figure 4 shows the correlations between UC and the results from prediction models based on NMR.

Bathen *et al.* used PLS to predict cholesterol and triglyceride in plasma and in the main lipoprotein fractions.[10] Relative prediction errors are calculated from their data and are shown in Table I. Their relative prediction errors are slightly higher than those found in our study, which is possibly due to the lower number of samples in the calibration (44 in that study compared to 72 in the study presented here) and the larger spectral range allowed in our study (Bathen *et al.* uses only the CH$_3$ and CH$_2$ peaks). Another explanation could be that in the present investigations diffusion edited NMR experiments are applied, suppressing the lactate signal which is usually found at 1.36 ppm on top of the CH_2 signal, possibly disturbing the models.

Table I   *Prediction results of lipoprotein subfractions (unit mg/dl) using UC as reference method*

		RMSECV[a]	# PC[b]	R[c]	Rel. RMSECV[d]	Bathen *et al.*
plasma	cholesterol	16.8	6	0.87	9 %	10 %
	triglyceride	12.2	3	0.99	9 %	11 %
VLDL	cholesterol	3.4	4	0.96	20 %	35 %
	triglyceride	10.3	4	0.98	12 %	19 %
VLDL 1	cholesterol	3.3	3	0.94	39 %	-
	triglyceride	11.7	3	0.97	23 %	-
VLDL 2	cholesterol	2.9	4	0.72	42 %	-
	triglyceride	6.4	3	0.64	39 %	-
IDL	cholesterol	2.6	5	0.71	31 %	46 %
	triglyceride	1.9	7	0.93	19 %	-
LDL	cholesterol	13.8	7	0.87	13 %	21 %
	triglyceride	3.7	7	0.75	17 %	29 %
LDL 1	cholesterol	4.7	7	0.70	24 %	-
	triglyceride	1.2	5	0.73	22 %	-
LDL 2-5	cholesterol	15.7	6	0.70	24 %	-
	triglyceride	3.0	5	0.51	26 %	-
LDL 6	cholesterol	5.1	3	0.78	31 %	-
	triglyceride	1.1	3	0.79	27 %	-
HDL	cholesterol	5.0	6	0.90	10 %	19 %
	triglyceride	2.5	8	0.92	18 %	19 %
HDL 1	cholesterol	3.4	7	0.85	29 %	-
	triglyceride	0.9	9	0.91	25 %	-
HDL 2	cholesterol	2.6	5	0.90	17 %	-
	triglyceride	1.1	7	0.90	25 %	-
HDL 3	cholesterol	3.9	6	0.54	17 %	-
	triglyceride	1.0	8	0.87	18 %	-

[a] Root Mean Square Error of Cross Validation (unit mg/dl)
[b] Number of PLS components used
[c] Correlation coefficient
[d] Relative RMSECV is calculated as RMSECV/(mean reference value)

To evaluate the prediction errors of our NMR spectroscopy-based PLS model, it is important to assess the error on the reference method. The error of the UC method was evaluated by dividing a blood sample into six parts and performing subfractionation on each of them, and the coefficient of variation between the six determinations was calculated. Furthermore, the week-to-week errors in the cholesterol and triglyceride measurements were taken into account, also calculated in terms of coefficients of variation. These errors add up to between 3 and 9 % for the UC determinations. Furthermore, an error in the VLDL subfractionation can be detected, where VLDL 1 and VLDL 2 do not add to total VLDL, but show a loss of up to 75 % – the average being 20 %. For HDL the error is much smaller, approximately ± 4 %. For LDL a similar error can be expected, but could not be directly assessed due to a dialysing step in the LDL subfraction preparation.

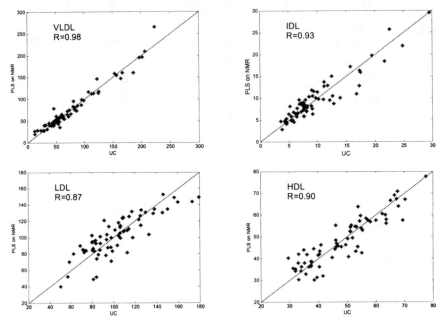

Figure 4   *Agreement between UC and NMR predictions on main fractions*

Despite the error in the UC results it is concluded that this error only accounts for a small part of the prediction errors of the PLS models, and since the spectral measurement is practically error-free, it would appear that there is considerable model error, the nature of which is being further investigated. One possible source of error could be the existence of non-bilinear data structure due to the continous density profile of lipoproteins.

Table II   *Description of clinical CHD risk criteria*

Criterion	Increasing CHD risk →			
Total ch (mg/dl)	< 200		200-240		> 240
Total tg (mg/dl)	< 150	150-199		200-500	> 500
LDL ch (mg/dl)	< 100	100-129	130-159	160-190	> 190
HDL ch (mg/dl)	> 60		60-40		< 40
Large VLDL tg (mg/dl)	< 7		7-27		> 27
Large HDL ch (mg/dl)	> 30		30-11		< 11
LDL particle no. (nmol/l)	< 1100	1100-1399	1400-1799	1800-2100	> 2100
LDL profile type (LDL1+2):(LDL5+6) ratio	> 1				< 1

In order to further evaluate the errors of the PLS prediction models, the 72 subjects were placed in risk groups based on clinical criteria often used in evaluating CHD risk. Table II shows a description of these criteria and the ranges of values describing each risk group (ranges adapted from www.liposcience.com). The "Large VLDL" is VLDL 1, the "Large HDL" is HDL 1 and the "LDL particle no." is calculated from the concentration of the protein apoB (prediction was performed, but not shown in Table I). The "LDL profile type" is calculated as LDL1 + LDL 2 divided by LDL 5 + LDL 6 (prediction results for LDL 2 and LDL 5 not shown in Table I) and is an alternative criterion to the common Type A/B classification based on the LDL particle size in nm, which was not determined in this study.

The assigned risk groups based on values predicted from NMR by PLS were compared to the assigned risk groups based on UC values for each subject. The number of subjects placed correctly and incorrectly as well as the agreement in percent can be seen in Table III.

Although the errors of the PLS models seem considerable for prediction of the subfractions, the determination of lipoproteins most important for evaluation of CHD risk are not affected, as the agreement between NMR predicted values and UC is between 76 and 100 %. The most important criteria for evaluation of CHD risk are the last four, of which only the large VLDL is not optimally predicted – probably due to the loss of lipoprotein during UC subfractionation. Accordingly, the evaluation of CHD risk is feasible from the NMR predicted values and gives risk assesments comparable to those obtained from ultracentrifugation, which themselves are subject to considerable error.

Table III *Placement in risk groups based on NMR predicted values compared to those based on UC values*

Criterion	Correctly placed	Incorrectly placed	Agreement
Total ch (mg/dl)	57	13	81 %
Total tg (mg/dl)	70	0	100 %
LDL ch (mg/dl)	54	15	78 %
HDL ch (mg/dl)	56	13	81 %
Large VLDL tg (mg/dl)	48	15	76 %
Large HDL ch (mg/dl)	58	10	85 %
LDL particle no. (nmol/l)	63	7	90 %
Small/dense LDL type	63	2	97 %

3.2 Lipoprotein subfraction quantification using GE as reference method

PLS models were also built between NMR data and the gel electrophoresis reference method. The obtained models were poor – the highest correlation coefficent being 0.71 for the VLDL model and only four of the 12 subfractions had a correlation coefficient higher than 0.5. This result clearly demonstrates that the physical phenomenon giving rise to different spectra for different lipoproteins in NMR is not the same as that used to separate the lipoproteins with GE.

Figure 5 shows correlations between UC and GE for the main fractions, VLDL, IDL, LDL and HDL. For GE, IDL, LDL and HDL are calculated as the sum of the subfractions.

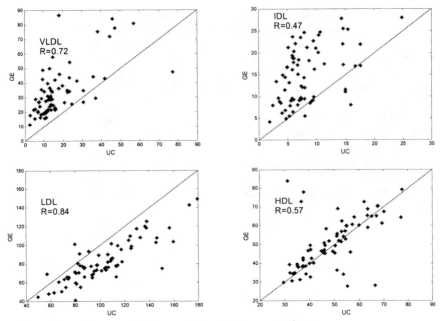

Figure 5   *Agreement between UC and GE on main fractions*

It is seen that the agreement between ultracentrifugation and GE is mediocre. VLDL and IDL are generally overestimated in GE, while LDL is underestimated. HDL is mostly quite well estimated, but with a few very severe outliers.

The correlation between GE and UC is quite good for LDL (R=0.84), indicating that this GE method safely can be used to monitor changes in LDL, while actual LDL concentrations determined by GE can not be directly compared with those obtained by UC.

Further work is being conducted to try to get a better correlation between GE and UC. One attempt is to change the boundaries between the subfractions in GE based on the ranges of subfractions from UC being measured by GE.

4 CONCLUSIONS

This work shows that the prediction of lipoprotein subfractions from high-resolution ^{1}H-NMR spectra is promising, although improvements in models leading to reduced prediction errors are required in order to make the method applicable in investigations where small changes in distinct lipoprotein subfractings are of interest.

The assessment of coronary heart disease risk is, however, feasible with the obtained prediction models.

Improvements can be expected from the following: 1) A larger number of samples will probably give better predictions due to better description of the underlying components representing the subfractions, 2) Measuring NMR spectra of plasma at 45°C has been reported to give markedly more signal from especially LDL, since part of the lipids are in an ordered state at 30°C, and thus not contributing to the NMR signal,

3) Optimisation of the data processing, for example the variable selection, with better PLS models as a goal could give better results.

In this respect a better understanding of the PLS model errors would be extremely helpful, especially if they are related to non-additive phenomena in the spectra of the lipoproteins.

References

1   W. J. Mack, R. M. Krauss and H. N. Hodis, *Arteriosclerosis and Thrombosis in Vascular Biology*, 1996, **16**, 697.

2   B. Lamarche, A. Tchernof, S. Moorjani, B. Cantin, G. R. Dagenais, P. J. Lupien and J-P. Després, *Circulation*, 1997, **95**, 69.

3   J. D. Otvos, 'Measurement of Lipoprotein Subclass Profiles by Nuclear Magnetic Resonance Spectroscopy' in *Handbook of Lipoprotein Testing*, eds., N. Rifai, G. R. Warnick and M. H. Dominiczak, AACC Press, Washington DC, 1997, Chapter 28, pp. 497-508.

4   N. Muñiz, D. Duncan and G. Neyer, The Frontiers in Lipoprotein and Vascular Disease, St. Louis, 2000.

5   P. Geladi and B. R. Kowalski, *Analytica Chimica Acta*, 1986, **185**, 1.

6   J. D. Otvos, E. J. Jeyarajah and D. W. Bennett, *Clinical Chemistry*, 1991, **37**, 377.

7   Y. Hiltunen, M. Ala-Korpela, J. Jokisaari, S. Eskelinen and K. Kiviniitty, *Magnetic Resonance in Medicine,* 1992, **26**, 89.

8   M. Ala-Korpela, Y. Hiltunen and J. D. Bell, *Anticancer Research*, 1996, **16**, 1473.

9   H. Serrai, L. Nadal, G. Leray, B. Leroy, B. Delplanque and J. D. de Certaines, *NMR in Biomedicine*, 1998, **11**, 273.

10 T. F. Bathen, J. Krane, T. Engan, K. S. Bjerve and D. Axelson, *NMR in Biomedicine*, 2000, **13**, 271.

11 J. D. Otvos, *Clinical Cardiology*, 1999, **22**, 21.

12 M. W. Baumstark, I. Frey, L. Berg and J. Keul, *Clinical Biochemistry*, 1992, **25**, 338.

13 M. J. Caslake and Packard, C. J., 'The Use of Ultracentrifugation for the Separation of Lipoproteins' in *Handbook of Lipoprotein Testing*, eds., N. Rifai, G. R. Warnick and M. H. Dominiczak, AACC Press, Washington DC, 1997, Chapter 29, pp. 509-529.

ANTIOXIDATIVE ACTIVITY OF SOME HERBS AND SPICES – A REVIEW OF ESR STUDIES

S.M. Đilas, J.M. Čanadanović-Brunet, G.S.Ćetković and V.T. Tumbas

Department of Organic Chemistry, University of Novi Sad, Faculty of Technology, Bulevar Cara Lazara 1, 21000 Novi Sad, Yugoslavia
E-mail: sdjilas@tehnol.ns.ac.yu

1 INTRODUCTION

In the early 1980s, consumers began looking for "natural" ingredients in their food.[1] This trend has promoted extensive research in the area of natural antioxidants.[2] Sources of natural antioxidant compounds are spices, herbs, tea, cocoa, oils, seeds, cereals, grains, fruits, vegetables, enzymes, proteins and protein hydrolysates.

Herbs have been always traditionally used, i.e. spices, for teas and as medicine. However, in the last ten years many steps, such as preclinical and clinical studies have been undertaken to give herbals and their preparations a firmer place in the range of products of pharmacies and drug stores. Herbals and especially herbal extracts are very attractive not only in the modern phytotherapy but also for food industry.[3,4]

Numerous investigations have proved that medicinal herbs contain compounds such as phenolic acids, derivates or isomers of flavones, isoflavones, flavonols, catechins, tocopherols, tannins, carotenoids, terpenoids, etc., which show antimicrobial, anthelminthic, anticarcinogenic, antineoplastic, antiviral, antiinflammatory, antiallergic, antioxidant and antiradical effect.[5,6]

Due to the revival of herbal extracts many questions arise, such as:

❖ what kind of herbal extracts can be used for the food industry?
❖ what technical data or physical prerequisites does a certain herbal extract need for being applicable in different types of food?
❖ which chemical and physicochemical measurements can be applied to evaluate their effect in food products?

The presence of free radicals is common in foods and arises from processing, including heating and radiation as well as oxidation during storage. Free radicals, and especially oxygen radicals as very reactive species, involve a lot of negative effects during lipid oxidation and damaging of protein and DNA. Because of that it is very important to define the interaction between free radicals and antioxidants in chemical and biological systems.

One approach to assessing antioxidant activity is to examine directly free radical production and its inhibition by an antioxidant. Electron spin resonance (ESR), also known as electron paramagnetic resonance (EPR) spectroscopy, has been used for decades to study free radicals and paramagnetic metals in chemicals and biological tissues, but its application to foods remains largely unexplored. However, many free radical species are highly reactive, with relatively short half-lives, and the concentrations found in natural systems are usually inadequate for direct detection by ESR spectroscopy. Spin-trapping is a chemical reaction that provides an approach to help overcome this problem. Using the spin trap, i.e. nitroso or nitrone compounds, it is possible to convert reactive free radicals to stable nitroxide radicals (adducts) with spectral hyperfine splittings that reflect the nature and structure of these radicals. The relative intensity of free radical formation can be determined because the ESR spectroscopy signal is directly related to the concentration of spin adducts. The height of the peaks in the spectra are proportional to the number of radical adduct molecules in the accumulating system. However, it is important to recognise that the ESR spectra of a particular spin adduct have unique characteristics that are dependent of the specific spin trap used and the free radical trapped, serving thus as sensitive and specific markers of the presence of a particular free radicals species.[7,8]

The chemical reactivity of plant extracts and individual organic compounds that react as antioxidants has been characterized using *in vitro* model systems based on the scavenging of reactive oxygen species or stable free radicals. Typical reactive oxygen species-generating systems employ UV light, radiation, pulse radiolysis, the hypoxanthine-xanthine oxidase system or metal ions (Fenton chemistry).[9] The substrate for oxidation in these reactive oxygen species-generating systems are frequently the lipids contained in tissue homogenates, liposomes, micelles, ghost cell membranes or simple and complex lipid systems such as methyl linoleate or low-density lipoprotein (LDL). Conjugated dienes and oxygen consumption are typical markers for defining the extent of oxidation. Stable free radicals that have been used to evaluate plant antioxidants include the radical cation of the compound 2,2'-azinobis(3-ethylbezothiazotline)-6-sulfonic acid (ABTS), the diphenyl-2-picrylhydrazyl radical (DPPH) and galvinoxyl (2,6-di-*tert*-butyl-α-(3,5-di-*tert*-butyl-4-oxo-2,5-cyclohexadien-1-ylidene)-p-tolyloxy) radical.[10,11]

This paper reviews investigations of antioxidant ability of some herbs and spices extracts to hydroxyl, peroxyl, DPPH, galvinoxyl or Fremy's radicals (potassium nitrodisulfonate), using a highly sensitive analytical method for detection of free radicals, electron spin resonance (ESR) spectroscopy.

The antioxidative activity of water, methanol, petroleum ether, chloroform, ethyl acetate or *n*-buthanol extracts of green and black tea, sage, oregano, rosemary, basil, winter and summer savory, hyssop, marjoram, gingko biloba, marigold, elecampane and wormwood extracts was assessed.

The paper also describes the chemical composition of phenolic secondary metabolites which are present in the investigated herbs and spices extracts. Finally, the mechanisms of antioxidative activity of these phenolic biomolecules are also summarised.

2 METHODS AND RESULTS

Strategies have been developed for measuring the antioxidative activity of different plant extracts as their ability to scavenge free radicals generated in the aqueous and lipophilic phases. The generation of a radical is coupled to oxidation of a substrate in which case measurement of the inhibitory effect of an antioxidant is based on the detection of either the radical or oxidation products.[12]

Madson et al.[13] used an ESR spin trapping technique to develop assays for evaluation of antioxidative activity of spices and spice extracts. They investigated the influence of water extracts of sage, oregano, rosemary, basil, winter and summer savory, hyssop and marjoram on the formation and transformation of hydroxyl radicals generated by the Fenton reaction and trapped by 5,5-dimethyl-1-pyrroline-N-oxide (DMPO). Oregano showed the highest antioxidative activity due to a variety of components including both hydrophilic and lipophilic compounds such as phenols and terpenes (Figure 1).

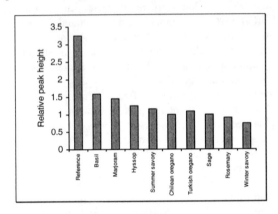

Figure 1 *Peak height of ESR spectra in arbitrary units for competition experiments between spin trap DMPO and spice extract after 5 min of reaction time. Reference: [DMPO]=9.8 mM, [Fe^{2+}]=4.9 µM and [H$_2$O$_2$]= 8.3 µM. All other samples as reference but with 0.042 mg/ml of spice extract added*

Noda et al.[14], using ESR spectroscopy and spin trapping methods (using DMPO as a spin trap), reported that the water-soluble extracts of gingko biloba and green tea had a strong hydroxyl radical scavenging activity. Thanks to this free radical scavenging ability gingko biloba extract is widely used for the treatment of various age-related disorders. ESR methodology has been used to assess the hydrogen-donating (antioxidant) ability of green and black tea to Fremy's radical and galvinoxyl radical in aqueous and organic solutions. Gardner et al.[15] estimated the abilities of the teas to reduce radicals in the aqueous phase by adding 3 ml of the tea (0.0025%) to an equal volume of a 1 mM solution of Fremy's radical. Antioxidants potential in an organic enviroment was assessed from the ability of tea extracts to reduce an equivalent volume of a 0.5 mM ethanolic solution of galvinoxyl radical. The obtained results showed that the green tea extract was a better radical scavenger (P<0.001) than black tea extract, reducing 20.8±1.0% and 24.2±2.0% more radicals in the ethanolic and aqueous systems, respectively. However, the amount of Fremy's radicals reduced by the extracts was less than the amount of galvinoxyl radicals, being only 60.2±1.5% and 58.5±0.9% for the green and black tea, respectively.

Elecampane (*Inula helenium* L.) has antibacterial, antifungal and antinflammatory activity, and it is used as a diuretic, carminative, sedative. To evaluate the antioxidative activity of elecampane, Čanadanović-Brunet et al.[16] used the DPPH free radical method. An ESR spectrum of stable DPPH free radicals is shown in Figure 2.

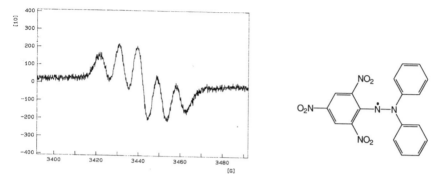

Figure 2 *ESR spectrum of stable the DPPH free radicals (0.4mM methanolic solution). The ESR spectra were recorded on an ESR Bruker 300E spectrometer (Rheinstetten, Germany) under the following conditions: field modulation 100 kHz, modulation amplitude 0.226 G, time constant 40.96 ms conversion time 671.089 ms, center field 3440 G, sweep width 100 G x-band frequency 9.64 GHz, microwave power 20 mW, temperature 23°C*

Hyperfine structure of the ESR spectra of DPPH free radicals is a result of interaction of the unpaired electron with two ^{14}N atoms (I=1) and consists of five lines of relative intensities 1:2:3:2:1. The hyperfine splitting constant is a_N=9.03G. The scavenging effect of petroleum ether, ethyl acetate and water extracts of elecampane on DPPH radicals is presented in Figure 3.

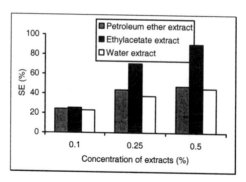

Figure 3 *Scavenging effect of different amounts of petroleum ether, ethyl acetate and water extracts of elecampane on DPPH radicals. The scavenging effect (SE) of extracts was defined as: $SE=100\% \cdot (h_0-h_x)/h_0$ where: h_0 - the height of the second peak in the ESR spectrum of DPPH radical of blank and h_x - the height of the second peak in the ESR spectrum of DPPH radical in reaction mixture with the addition of the extracts[16]*

The antioxidative activity of the investigated elecampane extracts is due to the hydrogen donor ability of the constituent biomolecules such as terpenes (alantolactone, germacren D-lactone, 8,9-epoxy-10-isobutyryl-oxythyimolisobutyrate, isoalantolactone) and polyphenols (quercetin-7-triglycoside).[17]

Sage (*Salvia officinalis* L.) is an excellent antiseptic and tonic. These activities of sage are mainly due to the presence of phenolic compounds (apigenin, apigenin-7-O-β-gly-coside, luteolin). Phenolic compounds isolated from methanolic extracts of sage function as reducing agents, free chain interrupters (scavenger effect), quenchers or inhibitors of the formation of singlet oxygen and inactivators of prooxidants metals.

Milić et. al.[18] studied the influence of chloroform, ethyl acetate and *n*-butanol extracts of sage on the formation and stabilization of hydroxyl radicals during Fenton reaction using ESR spectroscopy in combination with spin trapping method. The intensity of the ESR signal of DMPO spin adduct, DMPO-OH, (Figure 4), corresponding to the concentration of hydroxyl radicals (•OH), was reduced by all the investigated extracts. The 1:2:2:1 quartet of lines of the ESR spectrum with hyperfine coupling parameters (a_N and a_H=14.9 G) is typical for a DMPO-OH spin adduct.

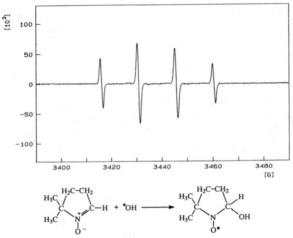

Figure 4 *ESR spectrum recorded 2.5 min after mixing of 1 ml 10 mM FeCl₂, 1 ml 80 mM DMPO and 1 ml 10 mM H₂O₂ on a Bruker ESR-300 E (Rheinstetten, Karlsruhe, Germany) spectrometer, set at the following conditions: receiver gain, 5·10⁵, time constant 81.92 ms, modulation amplitude 0.512 G, microwave power 20 mW, center-field 3440 G, conversion time 163.84 ms, temperature 23 °C[18]*

The investigated sage extracts possess antioxidative activity in the following order: ethyl acetate (89.78%), *n*-butanol (87.22%) and chloroform (53.58%).

In order to distinguish very reactive peroxyl radicals during lipid oxidation of edible oil, ESR spin trapping method can be also applied. Čanadanović-Brunet et al.[19] investigated the peroxyl radical scavenging activity of methanolic extract of sage (*Salvia officinalis* L.) during 600 minutes of catalytic oxidation of commercial sunflower oil. Using N-*tert*-butyl-α-phenylnitrone (PBN) as a spin trap it was possible to trap the obtained peroxyl radical (LOO•) and verify this by ESR (Figure 5). Hyperfine coupling parameters ($a_{N=}$14.75G and a_H=2.8 G) are typical for a PBN-OOL spin adduct.

This scavenging activity had the values between 13.5 and 26.8%, depending on the time of oxidation (Table 1).

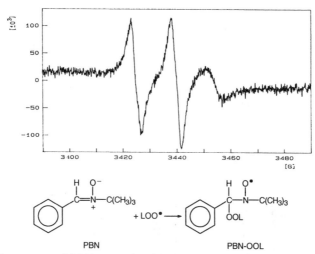

Figure 5 *ESR spectrum of PBN-peroxyl radical adduct formed during the reaction of 5 g of sunflower oil and 0.0213 g PBN. The ESR spectra were recorded on a ESR Bruker 300E spectrometer (Rheinstetten, Germany) under the following conditions: field modulation 100.000 kHz, modulation amplitude 0.204 G, time constant 327.68 ms conversion time 1310.72 ms, center field 3440.00 G, sweep width 100.00 G x-band frequency 9.64 gHz, power 20 mW, temperature 23°C[19]*

Table 1 *Peroxyl radical scavenging effect of the methanolic extract of sage during catalytic oxidation of sunflower oil*

Reaction time (min)	SE (%)
200	26.82
300	22.68
400	19.20
500	16.41
600	13.53

Reaction system: 5 g sample of sunflower oil, 48 mM PBN,
1 ml 0.1 mM $FeCl_2$ and 1 ml 1% methanolic extract of sage

The specific mode of inhibition is not clear but the individual flavonoids present in the methanolic extract of sage[20] may act as chelating agents, scavengers of lipid peroxyl radical acting as a chain breaking antioxidants, and the direct reaction between peroxyl and aroxyl radicals. The chloroform, ethyl acetate and *n*-butanol extracts of sage showed a smaller antioxidative effect on the formation and stabilization of peroxyl radicals during thermal oxidations of sunflower oil (60^0 C, 180 min).[21]

Beverages create a combination of innovation, health and enjoyment. Wormwood (*Arthemisia absinthium* L.) is used to produce bitter aromatic liqueurs. The plant is used in traditional medicine to stimulate the appetite, for gastrointenstinal complaints and as a choleretic, antiheminitic, etc. In our laboratory[22] we investigated the influence of different extracts of wormwood on the formation and stabilization of hydroxyl radicals during the Fenton reaction by ESR spectroscopy and spin trapping method using DMPO as a spin trap. Powdered dry plant material was macerated in 70% methanol for 72 h. After filtration, methanol was evaporated under the reduced pressure and the remaining water extract was fractioned using various organic solvents: petroleum ether chloroform, ethyl acetate and n-butanol. The intensity of the ESR signal was decreased by addition of all the investigated extracts (Figure 6). The following order of antioxidative activity has been established: ethyl acetate > methanol > *n*-butanol > petroleum ether > chloroform > water extract. Also, the investigation showed that the antioxidative activity increased with increasing the concentration of all extracts.

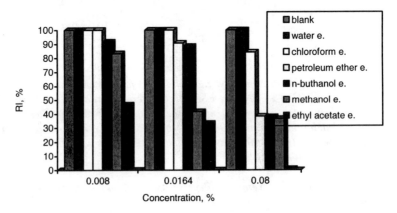

Figure 6 *The influence of different amounts wormwood extracts on the relative intensity of the ESR signal of the DMPO-hydroxyl radical spin adduct. The scavenging activity of the extracts was estimated by the percentage of decrease of the relative intensity (RI) of the signal of DMPO-hydroxyl radical adduct with reference to the control without extracts (blank, RI=100%)*[22]

In our country the yellow or golden-orange flowers of marigold (*Calendula officinalis* L.) are traditionally used as spice, tea and medicine. The active constituents in this plant include carotenoids, essential oils, flavonoids, sterols, tannins, saponins, triterpene alcohols, polysaccharides, bitter principle, mucilage and resin.[23]

The scavenging effect of the different concentrations of water and methanolic extracts of *Calendula officinalis* L. on hydroxyl radical produced by UV photolysis of hydrogen peroxide ($H_2O_2 \xrightarrow{\text{hv}} 2\ ^\bullet OH$) is shown in Table 2.[24]

Table 2 *Scavenging effect of the Calendula officinalis L. extracts on the hydroxyl radical produced by UV photolysis of hydrogen peroxide*

Extract concentration (%)	SE (%)	
	Water Extract	Methanolic Extract
0	0	0
0.065	54.93	51.61
0.075	67.93	66.67
0.095	86.89	84.67
0.125	90.47	88.93

The reactions were conducted by mixing 40×10^{-3} ml of 80 mM DMPO and 20×10^{-3} ml 20 mM H_2O_2 in a small Petri dish. Sample was exposed to the UV lamp set at $\lambda = 254$ nm for 10 min. The reaction mixture was then transferred to a quartz ESR flat cell ER-160-FC for ESR analyses.

The scavenging effect of aqueous and methanolic extracts of *Calendula officinalis* L. increased with increasing amounts of extracts. There is a small difference in the SE between water and methanolic extracts of *Calendula officinalis* L., both being good scavengers.

The amount of total flavonoids in marigold extract, prepared for the experiment, was 1.46 mg/g.[25] The results of TLC analyses show that the flavonoids, rutin and quercetin are present in water and methanolic extracts of *Calendula officinalis* L. According to the literature[26], these flavonoids have a main contribution to antioxidative activity of the investigated extracts. The mechanism of scavenging activity of the individual flavonoids is not clear but they may act as scavengers of hydroxyl radical by acting as a chain breacking antioxidants, as hydrogen donors with the formation of less reactive flavonoid (aroxyl-ArO) radical (ArOH + OH $\longrightarrow$ ArO$^{\bullet}$ + HOH) and as participant, in form of aroxyl radicals, in the reaction with hydroxyl radicals. A complex system of synergy effects and different symbiosis between certain substances (phenolic acids, flavonoids, tannins, terpenes and many others) are also very important to antioxidative activity.[27]

Different concentrations of aqueous extract of marigold also possess DPPH and $^{\bullet}$OH radical scavenging effect (Figure 7).[28,29] In this experimental procedure $^{\bullet}$OH radicals are produce during the Fenton reaction.

Water and methanol extracts of marigold, present at the concentration 0.125 %, showed a very good peroxyl radical scavenging ability during the catalytic oxidation of linoleic acid (reaction system: 3.2 mM linoleic acid, 48 mM PBN, 0.1 mM $FeCl_2$); reaction time: 24 h).[30] Scavenging effect of methanolic extracts was 42%, while the aqueous extract had a higher SE value (66.67%).

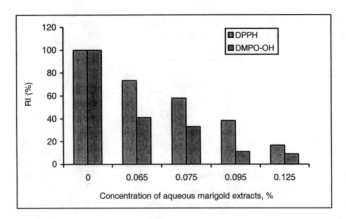

Figure 7 *The influence of different amounts of aqueous marigold extracts on the relative intensity of ESR signal of the DPPH free radical and DMPO-hydroxyl radical spin adduct*

Čanadanović-Brunet et al.[31] investigated the influence of the petroleum ether, chloroform, ethylacetate, n-butanol and water extracts on the thermal (60 °C, 24 h) oxidation of sunflower oil by ESR spectroscopy and spin-trapping method. The formation of the peroxyl radical (LOO$^\bullet$) in the oxidized sunflower oil in the presence of PBN, i.e. the PBN-peroxyl radical (PBN-OOL) spin adduct, was identified by analyzing the line intensity of ESR spectra. Based on the obtained results it can be concluded that all extracts inhibited the peroxyl radical formation in the following order: ethyl acetate > chloroform > petroleum ether > water > n-butanol (Figure 8).

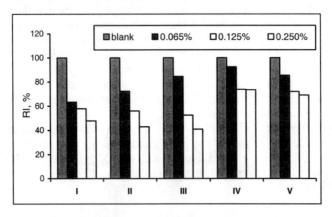

Figure 8 *The influence of petroleum ether (I), chloroform (II), ethyl acetate (III), n-butanol (IV) and water (V) extracts, concentration 0.065%, 0.125% and 0.25%, on intensity of the signal in the ESR spectrum of PBN-OOL spin adducts*[31]

3 CONCLUSION

Some generalization can be drawn from these results:

- ❖ ESR spectroscopy, especially in combination with spin trapping, can be employed to investigate the free radical processes in the model and real chemical and biological systems during some very complex oxido-reduction reaction.
- ❖ Some herb and spice extracts possess strong antioxidant ability and play an important role in transformation of hydroxyl, peroxyl, DPPH, galvinoxyl and Fremy's radicals.
- ❖ Secondary metabolites (derivatives or isomers of flavones, isoflavones, flavonols, catechins, tocopherols, tannins, carotenoids, terpenoids and many others, which are present in herbs and spices, contribute to the antioxidative activity.

References

1 O.I. Auroma, *J. Am. Oil. Chem. Soc.*, 1996, **73**, 1617.
2 M. Namiki, *Crit. Rev. Food Sci. Nutr.*,1990, **29**, 273.
3 B.Lj. Milić, S.M. Đilas, J.M. Čanadanović-Brunet and N.B. Milić, 'Radicali liberi in biologia, medicina e nutrizione' in *Farmacognosia*, eds. F. Capasso, R. De Pasquale, G. Grandolini and N. Mascolo, Springer, 2000, p. 449-462.
4 J. Lawless, *Illustrated Encyclopedia of Essential oils*, Barns and Noble, New York, 1999, part 1, p. 36.
5 P.J. Bryant and C.E. McQueen, *J. of Herbal Pharmacotherapy*, 2001, **1**, 17.
6 S.M. Đilas, J.M. Čanadanović-Brunet and G.S. Ćetković, *Chemical Industry*, 2002, **56**, 105.
7 D.J. Brackett, G. Wallies, M.F. Wilson and P.B. McCay, 'Spin Trapping and Electron Paramagnetic Resonance Spectroscopy' in *Free Radical and Antioxidant protocols*, eds., D. Armstrong, Humana Press Inc., Totowa, New Jersey, 1998, Part I, pp.15-27.
8 C.P. Poole, *ESR, A Comprehensive Treatise on Exeprimantal Techniques,* 2nd edn., John Wiley, New York, 1993, p.15.
9 B.Lj. Milić, S.M. Đilas and J.M. Čanadanović-Brunet, *Food Chem.*, 1998, **61**, 443.
10 W. Brand-Williams, M.E. Cuvelier and C. Berset., *Lebensm. Wiss. U. Tehnol.* 1995, **28**, 25.
11 E. Niki, *Chem.Phys. Lip.*, 1987, **44**, 227.
12 B.Lj. Milić, S.M. Đilas and J.M. Čanadanović-Brunet, *J. Mag. Res. Anal.*, 1996, **2**, 220.
13 H.L. Madsen, B.R. Nielsen, G. Bertelsen and L.H. Skibsted, *Food Chem.* 1996, **57**, 331.
14 Y. Noda, K. Anzia, A. Mori, M. Kohno, M. Shinmei and L. Packer, *Biochem. Molec. Biol. Intern.*, 1997, **42**, 35.
15 P.T. Gardner, D.B. McPhail and G.G. Duthie, *J. Sci. Agric.*, 1998, **76**, 257.
16 J.M. Čanadanović-Brunet, G.S. Ćetković and S.M. Đilas, *APTEFF*, 2002, **33**, in press.
17 F. Bohlman, P.K. Mahauta, J. Jakupovic, R.C. Rastogi and A.A. Natu, P*hytochem.*, 1978, **17**, 1165.
18 B.LJ. Milić, S.M. Đilas and J.M. Čanadanović-Brunet, *APTEFF*, 2000, **31**, 635.

19 J.M. Čanadanović-Brunet, S.M. Đilas, G.S. Ćetković and I.R. Nikačević, *APTEFF*, 2000, **31**, 3.

20 B.Lj. Milić, S.M. Đilas, J.M. Čanadanović-Brunet and M.B. Sakač, *Plant Phenolics*, University of Novi Sad, Faculty of Technology, p. 275

21 B.Lj. Milić, S.M. Đilas and J.M. Čanadanović-Brunet, Proceeding of 39. Conference "Production and Processing of oil seeds", Budva, June 1-6. 1998. p. 237.

22 S.M. Đilas, J.M. Čanadanović-Brunet and G.S. Ćetković, 2002, 6th International Conference on Application of Magnetic Resonance in Food Science, Paris, 4-6, September.

23 N.M. Dukić, 5th International Symposium of Interdisciplinary Regional Research, Szeged, October 4-6th, 2001, EP08.

24 S.M. Đilas, J.M. Čanadanović-Brunet and G.S. Ćetković, 5th International Symposium of Interdisciplinary Regional Research, Szeged, October 4-6th, 2001, EP05.

25 J.B. Harborne and T.J. Marby, *The Flavonoids Advances in Research*, Chapman and Hall, London, 1982, p.273.

26 A.R. Bilia, D. Salvini, G. Mazzi and F.F. Vincieri, *Chromatographia*, 2001, **53**, 210.

27 C.A. Rice-Evans, N.J. Miller and G. Paganga, *Free Radic. Biol. Med.*, 1996, **20**, 933.

28 S.M. Đilas, J.M. Čanadanović-Brunet, G.S. Ćetković and I.R. Nikačević, 1st International Symposium "Food in the 21st century", Subotica, November 14-17th, 2001, p. 373.

29 S.M. Đilas, J.M. Čanadanović-Brunet and G.S. Ćetković, IV Symposium "Modern Technologies and Economic Development", Leskovac, October 12-13th, 2000, p 18

30 S.M. Đilas, J.M. Čanadanović-Brunet, G.S. Ćetković and V.T. Tumbas, Production and Processing of Oil seeds, Proceedings of the 41st Oil Industry Conference, Budva, June 10-15th, 2002, p. 181.

31 J.M. Čanadanović-Brunet, S.M. Đilas, G.S. Cetkovic and V.T. Tumbas, Production and Processing of Oilseeds, Proceedings of the 43rd Oil Industry Conference, Budva, June 10-15th, 2002, p. 187.

THE SUBCELLULAR METABOLISM OF WATER AND ITS IMPLICATIONS FOR MAGNETIC RESONANCE IMAGE CONTRAST

S. Cerdán,[1] T. B. Rodrigues,[2] P. Ballesteros,[3] P. López[3] and E.P. Mayoral[3]

[1] Instituto Investigaciones Biomédicas "Alberto Sols" CSIC/UAM, c/ Arturo Duperier 4, E-28029 Madrid, Spain
[2] Present address: Department of Biochemistry, FCT University of Coimbra, 3001-4001 Coimbra, Portugal.
[3] Departamento de Química Orgánica y Biología, Facultad de Ciencias U.N.E.D., c/ Senda del Rey 9, E-28040 Madrid, Spain.

1 INTRODUCTION

Water plays a fundamental role in the maintenance of adequate physiology and metabolism in living organisms.[1] Under physiological conditions, the total water content of an average man weighting 70 Kg. is approximately 45 L. This volume is maintained constant by the operation of a plethora of endocrine mechanisms which control closely the balance between the supply of ca. 2,5 L. of drinking water per day and the elimination of the same volume through renal and intestinal excretion, evaporation through the respiratory tract and diffusion through the skin. Water turnover in humans varies with age and physiological status, being higher in infants than in adults and higher also in sport men than in sedentaries.[2] In healthy untrained adults, water molecules from dietary intake remain in the human organism for approximately ten to fifteen days. During this period, water serves not only as the solvent for physiological processes and the sink for all biological oxidations, but participates actively in vital biochemical reactions providing new hydrogen and oxygen atoms for exchange with those of existing tissue metabolites and macromolecules. These exchange processes involving water and the hydrogens from carbohydrates, lipids and proteins in tissues occur primarily at the cellular and subcellular levels. However, even though a large number of biochemical studies addressed earlier the subcellular compartmentation in the metabolism of carbohydrates, lipids and proteins, few reports have focussed on the subcellular metabolism of water.

Alterations in water homeostasis accompany most pathophysiological processes with evident manifestations such as cell swelling or edema in ischemic or traumatic episodes,[3] decreases in water rotational and translational dynamics during apoptotic responses,[4] dehydration events triggering dormancy and sporulation of yeasts and *Artemia*[5,6] or the increases in cellular dimensions preceding cellular division in bacteria.[7] Together, these circumstances reveal water metabolism and dynamics as crucial parameters for the diagnosis and follow up of many important pathologies as well as for the study of a variety of fundamental biological events. Interestingly, changes in water dynamics, metabolism and homeostasis are easily detected non invasively by Magnetic Resonance Imaging (MRI) methods, a relevant circumstance that confers MRI its enormous diagnostic potential. However, a better understanding on how water dynamics and

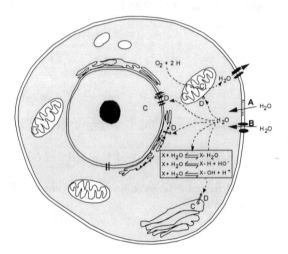

Figure 1. *A subcellular view of the metabolism of water. A: Diffusion through plasma membrane. B: Transport through water channels. C,D: Diffusion and transport to intracellular compartments. E: hydrogen and hydroxyl exchange. Reproduced from García-Martín et al.[8] with permission of the publisher.*

metabolism are translated into MR image contrast is needed before we can interpret unambiguously the information provided by MRI on physiological or pathological processes.

Figure 1 illustrates the main aspects of water metabolism at the cellular level.[8] Water turnover begins with the transport of water across the plasma membrane. This occurs mainly by diffusion through the phospholipid bilayer or through water channels named aquaporins. Once in the cytosol, water diffuses to intracellular organelles like the mitochondria, nucleus and endoplasmic reticulum where it is transported in a similar manner through the corresponding membranes. Subcellular water provides the liquid support for most biochemical reactions, contributing also to metabolism through a large variety of uncatalysed or enzyme catalysed exchanges which replace the hydrogens of macromolecules or metabolites with those of intracellular water. Finally, water molecules abandon the organelles and the cell through the same transport mechanism used to enter them, completing in this way the turnover of cellular water.

The following sections will describe in more detail the different aspects of water transport, diffusion and metabolism at the cellular and subcellular levels. The role of the different organelles and exchange events as determinants of intrinsic and extrinsic contrast in Magnetic Resonance Images will be discussed.

2 WATER TRANSPORT

Water transport is most frequently passive and reversible, occurring in the direction of the lowest water chemical potential. Thus, water can be transported inwards or outwards the cell through the same transport system, depending on the hypo-osmotic or hyper-osmotic condition of the extracellular medium. In this way, water fluxes through the plasma membrane determine the volume of cell, resulting in cell swelling or shrinking depending on the extracellular electrolyte and protein distributions. Transport of extracellular water to cellular interior is accomplished mainly by diffusion through the phospholipid bilayer, transport through water channels or co-transport with other metabolites.[9] The following sections analyse briefly these mechanisms.

2.1 Transport across the phospholipid bilayer

Diffusion through the phospholipid bilayer is a difficult process requiring relatively high activation energies (ca. 30-40 kJ mol[-1]) with permeabilities in the range 2-50 x 10^{-4} cm s[-1]. This process is thought to be coupled to the lateral diffusion of phospholipids (ca. 10^{-8} cm^2 s[-1]), a series of spontaneous translational events occurring in the liquid crystalline state above the transition temperature.[10] Under these conditions, a number of vacancies e xist within the phospholipid headgroups of the extracellular surface of the membrane which are able to accommodate temporarily molecules of water from the surrounding solvent. Headgroup dynamics cause then a specific headgroup to move into the vacancy, isolating the corresponding water molecule from the bulk solvent and pushing it down the bilayer. The fatty acid chains of membrane phospholipids, contain very frequently structural deffects (g-t-g kinks for example). The random translational dynamics of these deffects, force the isolated water molecule to move inside the membrane and progressively down the bilayer by additional movements. Eventually, the water molecule reaches the cytoplasmic side of the membrane, abandoning the bilayer inside the cell. The process is reversible, so intracellular water molecules may also diffuse back to the extracellular space using the same mechanism.

2.2 Transport through aquaporins

Transport through water channels in cells and organelles is more favourable than diffusion through the bilayer, with activation energies of 16-24 kJ mol[-1] and permeabilities of 200 x 10^{-4} cm s[-1]. Water transport through membrane pores was early anticipated in erythrocytes from the low values of the activation energy, similar to free diffusion.[11,12] The confirmation for this prediction was obtained soon by showing that water transport was sensitive to organomercurial reagents, which could affect critical sulfhydryl residues in the protein. Pioneering measurements on water transport were performed with radioactive isotopes in the fifties. However, more systematic studies involving NMR measurements could only start in the early seventies. Conlon and Outhred measured T_2 in eryhthrocyte suspensions doped w ith M n^{2+} a nd c alculated a r esidence t ime o f water in the erythrocyte (τ_r) of 9.6 ms.[13] This value was later refined by additional measurements at different field strengths and using T_1 sequences to a range of $9.8 < \tau_r < 14$ ms, similar to that previously found with radioactive isotopes.[14]

   Elucidadtion of the 3D structure of the water pores or aquaporins is a relatively recent achievement.[15] AQP1 is a tetramer, each monomer having six tilted, bilayer spanning α-helices which form a right handed bundle surrounding a central density. Up to ten different aquaporins have been described in mammals. Some of these transport exclusively water (AQP0, ACP1, AQP4, AQP5, AQP6 and AQP8). Others transport, in addition to water, other metabolites l ike g lycerol a nd u rea (AQP3, A QP7, A QP9). I n g eneral, a quaporins depict relatively similar 3D structures.

2.3 Co-transport

The contribution of water co-transport appears to be small in most cells, requiring significantly larger activation energies (78 kJ mol^{-1}) than free diffusion. Water has been reported to be cotransported through the K$^+$/Cl$^-$ co-transporter, the lactate monocarboxylate transporter and the Na$^+$/glucose transporter.[16,17] This type of water transport presents saturable kinetics and depends of the transmembrane concentration gradient of the co-transported solute, in contrast with diffusion, a not saturable process driven exclusively by the transmembrane difference of water chemical potential.

2.4 Active transport

Active transport of water is known to occur in leaky epithelia, where water is transported against its concentration gradient.[9] This requires the supply of sufficient energy to drive the thermodynamically this unfavourable process. Ion transport is thought to provide this energy although the coupling mechanisms remain uncertain.

3 INTRACELLULAR WATER DIFFUSION

Once water molecules have been transported to the cellular interior they diffuse through the cytoplasm and become further transported into the different subcellular organelles by similar mechanisms. Additional membrane transport processes and diffusion steps t ake place in the organelles, until the water molecules incorporated originally abandon the organelle and the cell. The following sections describe the translational and rotational diffusion of water and metabolites in the different subcellular organelles.

3.1 Cytoplasm

The a pparent t ranslational d iffusion c oefficient (ADC) o f w ater and metabolites in cells and tissues has been extensively studied *in vivo* using the Pulse-Field-Gradient-Spin-Echo (PFGSE) NMR technique.[9,18] Some examples of ADC values from water and metabolites determined using this approach in model solutions, rat erythrocytes, nucleated chicken erythrocytes and suspensions of rat liver mitochondria are given in Table 1.

The ADC value for water in the cytoplasm of rat erythrocytes is approximately one half from that of water in a model solution. Other cells show similar reductions, a finding that supports the general concept that translational diffusion of water in the cytoplasm is significantly slower than in model solutions and in the extracellular space. A large amount of information on cytoplasmic properties has been gathered by comparing the ADC's of molecules with different sizes and electrical charges.[8, 20-22]

3.2 Mitochondria

Water molecules exchange fast across the inner mitochondrial membrane with activation energy of 16-20 kJ mol^{-1} similar to those found for the free diffusion of water.[23] This suggests the presence of water channels in the mitochondrial membrane.

TABLE 1

ADC values (10^{-5} cm² s⁻¹, 25°C) for water, lactate, IMAC and ergothioneine in model solutions, suspensions of erythrocytes from rat or chicken and suspensions of rat liver mitochondria.

System / Metabolite	Model Solution[a] (ADC)	Rat Erythrocytes[b] (ADC$_i$)	Chicken Erythrocytes[c] (ADC$_i$)	Rat Liver Mitochondria[d] (ADC$_m$)
Water (R_G=0.755 Å)	1.997 ± 0.08 (n=21)	0.894 ± 0.12 (n=15)	0.900 ± 0.15 (n=5)	0.58 ± 0.19 (n=9)
Lactate (R_G=2.33 Å)	0.660 ± 0.12 (n=13)	0.121 ± 0.01 (n=13)	n.d.	n.d.
IMAC (R_G=3.44 Å)	0.670 ± 0.08 (n=15)	0.212 ± 0.04 (n=13)	0.219 ± 0.03 (n=8)	n.d.
Ergothioneine (R_G=4.52 Å)	0.460 ± 0.02 (n=4)	0.178 ± 0.06 (n=6)	0.167 ± 0.02 (n=13)	n.d.

Results are given as the mean ± s.e. of n measurements in different preparations. ADC: ADC in model solution, ADC$_i$: intracellular or intramitochondrial ADC. R_G: Molecular radius of gyration. [a] 0.2 M IMAC (Imidazol-1-ylacetic acid), 0.2 M lactate, 0.2 M ergothioneine, 10 mM TSP. [b] Rat erythrocyte suspensions (45% hematocrit). [c] Chicken erythrocyte suspensions (45% hematocrit). [d] Rat liver mitochondria suspension (100 mg protein/ml). Reproduced from García-Pérez et al.[19] with permission of the publisher.

Intramitochondrial rotations and translations of water have been investigated recently in suspensions of rat liver mitochondria. The intramitochondrial T_1 and T_2 of water are ca. 0.6 s and 0.006 s, respectively, being in both cases shorter than those of water in the cytoplasm. Similarly, water ADC in the mitochondria is approximately one half of that of the cytoplasm. Together, these results reveal that the translational and rotational motions of water in the intramitochondrial space are significantly slower than in the cytoplasm.

3.3 Nucleus

Translational diffusion of water in the nucleus has been investigated in nucleated chicken erythrocytes.[19] In this case, the attenuation of water and metabolite resonances was found to require smaller magnetic gradients than in non nucleated rat erythrocytes, suggesting faster ADC's. This difference was attributed to the presence of nucleus in chicken erythrocytes and allowed to calculate a value for the nuclear ADC of water of 2.0×10^{-5} cm² s⁻¹. This value is very similar to that found in the extracellular medium, suggesting that rotations and translations of water in the nuclear environment are not very different from those in the extracellular millieu (see Table 1). Similar results to those of nucleated chicken erythrocytes were obtained in neurons of *Aplysia Californica* and *Xenopus Laevis* oocytes.[24-26]

4 METABOLISM

Studies on the metabolism of water begun in the early thirties, soon after the deuterium and tritium isotopes of hydrogen were discovered.[27] Early studies, characterised two different classes of water in tissues.[28] The *exchangable water* could be removed by dessication while *non exchangable* water, constituted mainly by the hydrogens and hydroxyls of metabolites could only be released by combustion of the sample. The half-life of water in mice was estimated as three days. However, the metabolism of water at the cellular and subcellular levels remained insufficiently explored. The following sections address the basic mechanisms of hydrogen exchange and describe recent advances in our understanding of the metabolism of water and its implications for MRI contrast.

4.1 Non enzymatic hydrogen exchange

The most frequent exchange reactions between intracellular water molecules are probably those involving intermolecular proton or hydroxyl transfers

$$H_2O + H_3O^+ \quad \rightarrow \qquad H_3O^+ + H_2O \qquad [1]$$

$$H_2O + OH^- \quad \rightarrow \quad OH^- + H_2O \qquad [2]$$

The rate constants for these two processes have been determined by ^{17}O NMR as 8.2 $\times 10^9$ and 4.6 $\times 10^9$ $M^{-1}s^{-1}$, corresponding to activation energies of 10.9 and 11.3 kJ mol^{-1}, respectively.[29] These activation energies are smaller than the activation energy for the self diffusion of water, indicating that proton and hydroxyl transfers are limited by diffusion.

Another important class of hydrogen exchange reactions include prototropic transfers to nucleophyles (:B) of the type;

$$H_3O^+ + :B \rightarrow H_2O + HB^+ \qquad [3]$$

These reactions take place most normally between water and the OH or NH groups of biomolecules, with similar kinetics and energetics than the proton transfers between two water molecules. The process is acid or base catalysed and diffusion controlled as indicated above. The forward rate constant k_{tr} can be defined as

$$k_{tr} = k_D \,[10^{\Delta pKa}/(10^{\Delta pKa} +1)] \qquad [4]$$

where the k_D value is approximatelly 10^{10} M^{-1} s^{-1} and ΔpK_a represents the difference between the pK_a's of the proton donor and acceptor molecules.[30,31]

Keto-enol tautomerism constitutes and additional and frequent way of incorporation of solvent hydrogens into metabolites. The tautomerism causes hydrogens adjacent to carbonyl groups to be significantly more acidic than those of hydroxyl groups.[32]

In general, non enzymatic acid or base catalysed hydrogen exchange occurs very slowly at neutral pH, making hydrogen turnover in metabolites to be dominated by enzyme catalysed reactions under physiological conditions.

4.2 Enzyme catalysed hydrogen exchange

Table 2

Enzyme catalysed ^2H-^1H exchange reactions involved in the deuterium substitution of specific hydrogens from some relevant metabolites.

Metabolite Hydrogen Replaced by ^2H	Enzyme(s) or Pathway
Aspartate H3 *pro-R*	Fumarase
Aspartate H3 *pro-S*	TCA[a]
Aspartate H2	Aspartate aminotransferase
Glutamate H4,H4′	Citrate synthase, TCA[b]
Glutamate H3 *pro R*	Aconitase
Glutamate H3 *pro S*	Isocitrate dehydrogenase
Glutamate H2	Aspartate aminotransferase
	Glutamate dehydrogenase
Alanine H3	Alanine aminotransferase
	TCA[b], Pyruvate Kinase
Alanine H2	Alanine aminotransferase
Lactate H3	TCA[b]
	Pyruvate kinase
Lactate H2	Aldolase, TIM[c], GAPDH[c]
Glucose H6	Fumarase
Glucose H5	Enolase
Glucose H4	Aldolase, TIM, GAPDH
Glucose H3	Aldolase, TIM, GAPDH
Gucose H2	Glucose 6P isomerase
Glucose H1	Fumarase, Glucose 6P isomerase

[a] Derived from $(4-^2H)$ α-ketoglutarate and acetyl- CoA.
[b] Derived from $(3-^2H)$ pyruvate and (2-2H) acetyl-CoA. originated from $(3-^2H)$ malate or oxalacetate through malic enzyme or phosphoenolpyruvatecarboxykinase and pyruvate kinase activities.
[c] TIM: triosephosphate isomerase, GAPDH: glyceraldehyde 3P dehydrogenase

Table 2 summarises the most relevant enzyme catalysed ^1H-^2H exchange reactions in metabolites as detected by ^1H, ^2H and ^{13}C NMR.[33] Hydrogen turnover in the tricarboxylic acid cycle and the aminotransferase reactions are described in detail in the next sections. Additional information on these exchanges and the remaining hydrogen exchange pathways including redox processes, glycolysis, gluconeogenesis and fatty acid synthesis and degradation may be found in the classical monograph or Walsh[34] and in refs. 8, 33 and 35.

4.2.1. The tricarboxylic acid cycle. Meticulous ^3H and ^2H studies established the mechanisms o f h ydrogen i ncorporation a nd r emoval d uring t he s tereospecific h ydration-dehydration and red-ox reactions of the tricarboxylic acid cycle (Figure 2).[36] Following the carbon pathway, the C1, C2 and C3 carbons or the C4 and C5 carbons of α–ketoglutarate/glutamate are derived from the C4, C3 and C2 carbons of oxalacetate or the C2 and C1 carbons of incoming acetyl-CoA, respectively. Considering the hydrogen pathway and beginning with fumarate, *trans* addition of ^2H_2O to its double bond results in (3R, 3-^{2}H) malate first and (3R, 3-^{2}H) oxalacetate subsequently. Oxalacetate condenses then with acetyl-CoA to originate citrate which is metabolized to isocitrate and α-ketoglutarate. The original 3R deuteron of malate is lost while forming the C2 carbonyl of α-ketoglutarate. In contrast, the H3$_{proR}$ and H3$_{proS}$ hydrogens from α-ketoglutarate are derived from the solvent and incorporated through the activities of aconitase and isocitrate dehydrogenase over the carbon originally derived from oxalacetate C2 (H in circle and triangle on α-ketoglutarate in Figure 2). The H4 hydrogens from α-ketoglutarate are derived from the methyl group of acetyl CoA and may become labelled if the citrate synthase reaction occurs in ^2H_2O. In the next step, α-ketoglutarate is decarboxylated to succinate. In this process, the C3 and C4 α-ketoglutarate carbons (or the H3 and H4 hydrogens) become the C2 and C3 carbons (or the H2 and H3 hydrogens) of succinate, respectively. Subsequently, succinate dehydrogenase removes in *trans*, one H2 and one H3 hydrogen from succinate to yield fumarate and begin a new turn of the cycle. Interestingly, while the carbons of α-ketoglutarate/glutamate remain in the cycle for more than one turn, all hydrogens from α-ketoglutarate/glutamate are renovated by solvent hydrogens during every turn of the cycle.

4.2.2 Aminotransferase reactions. Aspartate and alanine aminotransferases are known to incorporate deuterons from the solvent into the C2 position of the amino acids through a mechanism involving typical pyridoxal phosphate catalysis.[37-39]

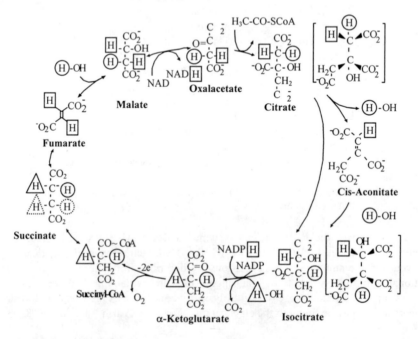

Figure 2. *Hydrogen incorporation and removal in the tricarboxylic acid cycle.*

Classical experiments monitored by ^{1}H NMR the aspartate and alanine aminotransferase reactions in ^2H_2O containing medium. In the case of aspartate aminotransferase, deuterons were incorporated into the C2 carbon of glutamate and aspartate only. Notably, even though the H3 hydrogens of glutamate and aspartate are not exchanged with the solvent by the transaminases, theirsubstitutions have important effects on subsiquent transaminations.[40] Substitution of the H3$_{pro-S}$ inhibits deuteration of H2 while susbstitution of H3$_{pro-R}$ allows H2 deuteration reaction to proceed normally. Alanine aminotransferase is known to incorporate solvent deuterons both in H2 and in H3 alanine.[37]

4 HYDROGEN TURNOVER AS DETECTED BY (^{2}H,^{1}H) ^{13}C NMR

An important limitation to progress in the understanding of hydrogen turnover in tissue metabolites has been the lack of a suitable technique to observe this phenomenon. Recently (^{2}H, ^{1}H) ^{13}C NMR spectroscopy has been shown to have important potential in this respect.[33,35] The general strategy is based on the indirect detection of ^{2}H incorporation into a ^{13}C labeled metabolite through its effects on the J couplings and isotopic shifts of the attached geminal and vicinal ^{13}C carbon resonances (Figure 3).[8,41] Replacement of one geminal hydrogen by one deuteron splits the original ^{13}C singlet (s) into a 1:1:1 triplet (t) ($19.21 < {}^1J_{C2H} < 22$ Hz) inducing a geminal isotopic shift ($-0.25 < \Delta_1 < -0.33$ ppm).

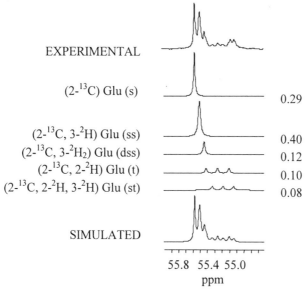

Figure 3. *Deconvolution of the ($^1H,^2H$) $^{13}C2$ resonance of (2-^{13}C) glutamate into (^1H) ^{13}C and (^2H) ^{13}C components. ^{13}C NMR spectra (150.13 MHz) were acquired from a representative extract of mouse liver perfused (15 min) with 6 mM (3-^{13}C) alanine in buffer containing 50% v/v ^2H_2O. s: singlet, ss: shifted singlet, dss: doubly shifted singlet, t: triplet, st: shifted triple. Numbers on the left indicate fractional contribution. Reproduced from García-Martín et al.[35] with permission of the publisher.*

Multiple replacements of geminal hydrogens with two or three deuterons result in additive isotopic shifts and ^{2}H-^{13}C coupling patterns of five or seven line multiplets, respectively. Vicinal deuterium substitutions cause smaller isotopic shifs (-0.03 < Δ_2 < -0.11 ppm), remaining as shifted singlets (ss) since vicinal ^{13}C,^{2}H couplings are too small to be resolved. Vicinal isotopic shifts are also additive and a double vicinal deuterium substitution results in a doubly shifted singlet (dss). Therefore, careful analysis of the shifted and unshifted (^{2}H,^{1}H) ^{13}C resonances of a deuterated ^{13}C isotopomer, allows the determination of the number of deuterium replacements, their relative contributions and the geminal or vicinal location with respect to the observed ^{13}C carbon as illustrated in Figure 3 with (2-^{13}C) glutamate.

Deconvolutions of the ^{13}C,^{2}H multiplets of the C2 glutamate resonance of mouse liver extracts prepared after increasing perfusion times with (3-^{13}C) alanine and buffer containing 50% ^{2}H$_2$O have allowed recently the determination of the time courses of production of the corresponding ^{13}C-^{2}H isotopomers (Figure 4). The kinetics of ^{1}H-^{2}H exchange in the H2 and H3 hydrogens of hepatic glutamate contain unique information on intracellular glutamate trafficking. Even though ^{1}H-^{2}H exchange is an ubiquitous cellular process occurring both in cytosol and mytochondria, the slow exchanges due to the tricarboxylic acid cycle occur exclusively in the mitochondrial space, allowing to distinguish between both intracellular environments on the basis of the different kinetics of deuteration of the glutamate H2 and H3 hydrogens (Figure 4). Interestingly, the H2 and H3 hydrogens of glutamate depict two consecutive ^{2}H exchanges. This means that the deuteron incorporated first must be removed, before the second deuteron occupies the same position at a later stage.

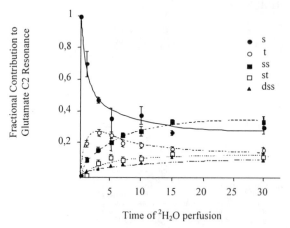

Time of ^{2}H$_2$O perfusion

Figure 4. *Kinetics of deuteration of the H2 and H3 hydrogens from (2-^{13}C) glutamate. Similar deconvolutions to those of Figure 4 were performed in extracts from mouse liver prepared after increasing times of perfusion with (3-^{13}C) alanine in buffer containing 50% v/v ^{2}H$_2$O. The simbols correspond to the shifted and unshifted resonances of Figure 4. s: singlet, ss: shifted singlet, t: triplet, st: shifted triplet, dss: doubly shifted singlet. Lines indicate the best fit to a minimal model of hydrogen exchange. Reproduced from García-Martín et al.[35] with permission of the publisher.*

In the case of glutamate H2, (2-^{2}H, 2-^{13}C) glutamate (t in Figures 3 and 4) is generated first, followed later by a slower production of (2-^{2}H, 3-^{2}H, 2-^{13}C) glutamate (st in Figures 3 and 4). In the case of H3, (3-^{2}H, 2-^{13}C) glutamate (ss in Figures 4 and 5) depicts faster kinetics than (3,3'-^{2}H$_2$, 2-^{13}C) (dss in Figures 3 and 4).

This sequence of deuteration events agrees well with the mechanism of α-ketoglutarate/glutamate exchange between mitochondria and cytosol predicted in the malate-aspartate shuttle (Figure 5). First, preexisting cytosolic (2-^{13}C) α-ketoglutarate molecules formed during a preperfusion period without ^{2}H$_2$O, become rapidly deuterated in H2 by cytosolic aspartate aminotransferase after changing the perfusion medium to ^{2}H$_2$O containing buffer. This yields the fast (2-^{2}H, 2-^{13}C) glutamate component (t and f labels in panel 5A). Similarly, the fast (f) component of (3-^{2}H, 2-^{13}C) glutamate (ss) represents most probably the activity of cytosolic NADP-isocitrate dehydrogenase on preexisting isocitrate molecules (ss and f in panel 5A). Later, (2-^{13}C) glutamate molecules deuterated in the cytosol enter the mitochondria (Figure 5, process a), where the cytosolic H2 deuteron is lost by transamination to mitochondrial (2-^{13}C) α-ketoglutarate (Figure 5, p rocess b). The remaining hydrogens or deuterons from

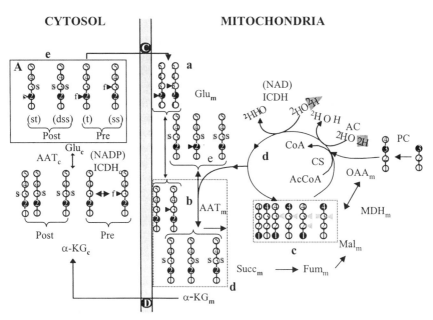

Figure 5. *M echanism o f α- ketoglutarate/glutamate e xchange b etween m itochondria a nd cytosol as detected by the kinetics of deuteration in the H2 and H3 hydrogens of glutamate. A: (^{2}H^{13}C) glutamate isotopomers detected in Figures 3 and 4. Pre: before TCA cycle. Post: after TCA cycle. f: fast deuteration, s: slow deuteration .^{13}C or ^{12}C, black or white circle. ^{2}H, grey triangle or square. Subscripts c or m refer to cytosolic or mitochondrial. AAT: aspartate aminotransferase, ICDH: isocitrate dehydrogenase, CS: citrate synthase, PC: pyruvate carboxylase, AC: aconitase. C: glutamate aspartate exchanger, D: dicarboxylate carrier. Reproduced from García-Martín et al.[35] with permission of the publisher.*

mitochondrial $(2-^{13}C)$ α-ketoglutarate are lost subsequently during the next turn of the cycle. During TCA cycle metabolism, $(2-^{13}C)$ α-ketoglutarate molecules derived from the cytosol originate an equimolar amount of $(1-^{13}C)$ and $(4-^{13}C)$ succinate molecules loosing also the cytosolic ^{13}C label in C2 α-ketoglutarate/glutamate (Figure 5, process c). However, new $(2-^{13}C)$ α-ketoglutarate molecules are continuously being produced *de novo* by carboxylation and metabolism of $(3-^{13}C)$ alanine in the tricarboxylic acid cycle (Figure 5, process d). The newly formed molecules of $(2-^{13}C)$ α-ketoglutarate carry $^2H3$ or $^2H3'$ labels or $^2H3$ and $^2H3'$ labels (Figure 5, grey triangles and squares) derived from the successive hydration-dehydration reactions of aconitase and NAD-isocitrate dehydrogenase, respectively (Figure 5, process e). New $(2-^{13}C, 3-^2H)$ and $(2-^{13}C, 3,3'-^2H_2)$ α-ketoglutarate molecules formed in the cycle, transaminate soon to originate $(2-^{13}C, 3-^2H, 2-^2H)$ and $(2-^{13}C, 3,3'-^2H_2)$ glutamate molecules (st, dss in Figures 3 and 4). However, the rate of trasamination of the newly formed $(2-^{13}C)$ α-ketoglutarate molecules in the cycle is much slower (Figure 5 s labels) than that of preexisting cytosolic $(2-^{13}C)$ α-ketoglutarate, since it is necessarily limited by the slow TCA cycle.

5 IMPLICATIONS FOR MAGNETIC RESONANCE IMAGING CONTRAST

MRI contrast arises mainly from intrinsic differences in water content and relaxation rates of the water protons in the different tissues.[42] Since differences in water content are relatively small, most of the contrast is derived from differences in the tissue relaxation rates. These depend ultimately on the rates water of relaxation in the corresponding extracellular and vascular spaces, in the intracellular space and on the various exchanges occurring within these different environments. The information provided above allows to evaluate more closely these processes and provide a hierarchy for their contribution to image contrast using some approximations. Considering only the intra-and extracellular spaces and a fast exchange of water between them, the following expression holds

$$1/T_{1\ tissue} = V_i/T_{1i} + V_e/T_{1e} \qquad\qquad [5]$$

where $1/T_{1\ tissue}$ is the observed proton relaxation rate of the tissue; V_i or V_e represent the tissular intra- or extracellular volumes and T_{1i} or T_{1e} the intra- and extracellular longitudinal relaxation times, respectively. Under physiological conditions, the extracellular space, including the vascular and interstitial spaces, can reach up to 35 % of the tissular volume while the intracellular space occupies the remaining 65%.[1] Thus, for 1g of representative tissue with typical intracellular T_1's of the order of 0.8 s and extracellular T_1's ca. 2.0 s, expression [7] becomes $1/T_{1\ tissue} = 0.8\ s^{-1}\ g^{-1} + 0.2\ s^{-1}\ g^{-1}$, indicating that logitudinal relaxation rates in tissues contain approximately an 80% contribution from the intracellular space. A different situation arises when an extracellular paramagnetic contrast agent is used. For conventional contrast agents[43] as Gd(III)DTPA or Gd(III)DOTA with relaxivities of the order of 4-5 $s^{-1}\ mM^{-1}$, the presence of 2-3 mM paramagnetic agent would increase up to 10-15 times the relaxivity of the extracellular compartment, making then extracellular relaxation rates to dominate expression [5].

Similar expressions to [5] can be written for transversal relaxation times T_2 and ADC values. For intra- and extracellular T_2's of 0.12 s and 2.0 s, the intracellular compartment contributes 97% of the observed transversal relaxation. Similarly, for intra- and extracellular ADC's of $0.8\ 10^{-5}\ cm^2\ s^{-1}$ and $1.8\ 10^{-5}\ cm^2\ s^{-1}$, the intracellular compartment contributes approximately 86% of the diffusion weighting. Therefore, T_2 weighted images and diffusion weighted images are dominated also by the contributions of the intracellular compartment under physiological conditions. A different situation arises when superparamagnetic or ferromagnetic particles are used as T_2 contrast agents. These agents can increase by an order of magnitude or more the transversal relaxivity of the vascular space which may dominate then the transversal relaxation rate of the tissue.

It is possible to progress one step further on the physiological mechanisms of intracellular relaxation by evaluating the relative contributions of different organelles to overall cellular relaxation. The intracellular longitudinal and transversal relaxations or the ADC's (termed in general M_i), can be expressed as the weighed sum of the corresponding contributions from the cytoplasmic, mitochondrial and nuclear compartments

$$1/M_i = V_c/ M_c + V_m/M_m + V_n/M_n \qquad [6]$$

where M_c, M_m and M_n represent cytosolic, mitochondrial and nuclear values of T_1, T_2 or ADC and V_c, V_m and V_n represent the relative volumes of cytoplasm, mitochondria and nucleus, respectively. Taking the liver parenchima as an example,[44] V_c, V_m and V_n have approximate values of 58 %, 28 % and 8 %; T_{1c}, T_{1m} and T_{1n} are estimated ca. 0.8 s, 0.6 s and 1.8 s; and T_{2c}, T_{2m} and T_{2n} are in the range 0.12s, 0.006s and 1.8s, respectively. Therefore, according to [6] cytosol, mitochondria and nucleus contribute approximately 59 %, 38 % and up to 3 % of the observed longitudinal relaxation rates; 9%, 90% and 1% of the observed transversal relaxation rates and 56%, 40% and 4% of the observed ADC weighing in liver parenchima, respectively. It is very interesting to note here that while intracellular T_1's and ADC's reflect the cytosolic contributions, intracellular T_2's appear to be mainly dominated by the mitochondrial contributions. Thus, T_1 and ADC weighted images reflect mainly the properties of the cytoplasm, while T_2 weighed images reveal the properties of the mitochondrial environment. In becomes then possible to emphasise in a fully non invasive manner, cytoplasmic or mitochondrial events by MRI simply by choosing the pulsing conditions to enhance T_1 or T_2 weighing.

Finally, the results described here allow to compare the relative contributions of hydrogen turnover in metabolites and proteins as potential sources of tissue contrast in MRI. One way to obtain this information is to compare the number and dynamics of water or hydrogen exchanges originated in metabolites and proteins contained in 1 g of tissue. Metabolites can reach an approximate concentration of 50 $\mu mol.g^{-1}$, while proteins can achieve 150 mg g^{-1}. For an average protein size of 80 kD, this yields an average concentration of 1-2 $\mu mol\ g^{-1}$. The number and frequency of hydrogen exchange in metabolites is much smaller than in proteins. Metabolites contain an average of 4 sites of hydrogen exchange with an average residence times of 180 s for the fastest processes. On the other hand, proteins may induce relaxation in solvent water either by (i) the exchange of a small number of tightly bound water molecules (ca 5-15) with residence times ranging 1 ns to several ms (ii) the exchange of very large number of hydration molecules (>1000) of water exchanging in the ns or subns range or (iii) through hydrogen transfers from the many OH or NH groups (>200) in amino acids to the solvent with residence times in the ns to ms scale at physiological pH.[45,46] Therefore the number and frequency of hydrogen exchange processes in proteins and other macromolecules appears to dominate contrast in MR images of cells and tissues.

6 CONCLUDING REMARKS

In summary, we provided an overview of the subcellular water metabolism and its potential implications for MRI contrast. Assuming fast exchange conditions, the values of T_1, T_2 and ADC of subcellular organelles determine the m agnetic p roperties o f t issues a s n ormally detected in MRI scans. Notably, cytoplasmic contributions dominate T_1 and ADC weighted images, while mitochondrial contributions dominate the T_2 weighted contrast. Concerning MRI contrast at the molecular level, enzyme catalysed hydrogen exchange does not appear to contribute appreciably to MRI contrast which seems to be derived mainly from uncatalysed water and hydrogen exchanges between macromolecular components and the intracellular solvent.

7 ACKNOWLEDGEMENTS

This work has been made possible by a strategic group grant from the Community of Madrid to P.B. and grant SAF 2001-2245 from the Ministry of Science and Technology to S.C.

8 BIBLIOGRAPHY

1.   A. Guyton, *Textbook of Medical Physiology*; 8th Edn., W.B. Saunders Co., Philadelphia, 1991, pp 274
2.   J. B. Leiper, A. Carnie and R. J. Maughan, *Br J Sports Med.,* 1996, **30**, 24.
3.   H. K. Kimelberg, *J Neurotrauma,* 1992, **9** *Suppl 1*, S71.
4.   S. Hortelano, M. L. Garcia-Martin, S. Cerdan, A. Castrillo, A. M . A lvarez a nd L. Boscá *Cell Death Diffe,r* 2001, **8**, 1022.
5.   J. S. Clegg and J. Cavagnaro, *J Cell Physiol* 1976, **88**, 159.
6.   J. S. Clegg, *J Exp Biol,* 1974, **61**, 291.
7.   N. Grover and C. Woldringh, *Microbiology* 2001, **147**, 171.
8.   M. L. García-Martín, P. Ballesteros and S. Cerdán, *Progr. NMR Spec.* 2001, **39**, 41.
9.   T. Zeuthen, *Molecular Mechanisms of Water Transport*; Springer: New York, 1996.
10. T. H. Haines, *FEBS Letters* 1994, **346**, 115.
11. A. S. Verkman, *J Membr Biol*, 2000, **173**, 73.
12. M. Borgnia, S. Nielsen, A. Engel and P. Agre, *Annu. Rev. Biochem.* 1999, **68**, 425.
13. T. Conlon and R. Outhred, *Biochim. Biophys. Acta*,1972, **288**, 354.
14. M. D. Herbst and J. H. Goldstein, *Am. J. Physiol.* 1989, **256**, C1097.
15. A. Engel, Y. Fujiyoshi and P. Agre, *Embo J.,* 2000, **19**, 800-6.
16. T. Zeuthen, *J. Gen. Physiol.* 1994, **478**, 203.
17. T. Zeuthen and W. D. Stein, *J. Membr. Biol.* 1994, **137**, 179.
18. A. R. Waldek, P. Kuchel, A. J. Lenon and B. E. Chapman, *Prog. NMR Spec.* 1997, **30**, 39.
19. A. I. García-Pérez, E. A. López-Beltrán, P. Klüner, J. Luque, P . B allesteros a nd S . Cerdán, *Arch. Biochem. Biophys.* 1999, **362**, 329.
20. K. Luby-Phelps, D. L. Taylor and F. Lanni, *J Cell Biol* 1986, **102**, 2015.
21. K. Luby-Phelps, P. E. Castle, D. L. Taylor and F. Lanni, *Proc Natl Acad Sci U S A* 1987, **84**, 4910.

22. K. Luby-Phelps, *Int Rev Cytol* 2000, **192**, 189.
23. E. A. López-Beltrán, M. J. Maté and S. Cerdán, S., *J. Biol. Chem.* 1996, **271**, 10648.
24. N. R. Aiken, E. W. Hsu, A. Horsman and S. J. Blackband, *Am. J. Physiol.* 1996, **271**, C1295.
25. S. J. Blackband and M. K. Stoskopf, *Magn. Res. Imaging*, 1990, **8**, 191.
26. S. J. Blackband, D. L. Buckley, J. D. Bui and M. I. Phillip, *Magma* 1999, **9**, 112.
27. G. N. Lewis, *Science* 1934, **79**, 151.
28. R. Schoenheimer and D. Rittenberg, *Physiol. Rev.* 1940, **20**, 218.
29. J. A. Glasel, in *Water a comprehensive treatise*; Franks, F., ed., Plenum Press: New York, 1972; vol. 1, pp 215.
30. M. Eigen, *Angew. Chem. (Intl. Ed.)* 1964, **3**, 1-19.
31. K. Wütrich, *NMR of Proteins and nucleic acids*; Wiley-Interscience Publication: New York, 1986.
32. M . B. Smith and J. March, *Advanced Organic Chemistry; Reactions, Mechanisms and Structure*, 5th Edn., John Wiley and Sons: New York, 2001.
33. M. Moldes, S. Cerdán, P. Erhard and J. Seelig, *NMR Biomed.* 1994, **7**, 249.
34. C. Walsh, *Enzymatic Reaction Mechanisms*; W.H. Freeman and Co.,San Francisco, 1979.
35. M. L. García-Martín, M. A. García-Espinosa, P. Ballesteros, M. Bruix and S. Cerdán, *J. Biol. Chem.* 2002, **277**, 7789.
36. H. R. Mahler and E. H. Cordes, in *Biological Chemistry*, Harper International: New York, 1967, p 525.
37. A. J. L. Cooper, *J. Biol. Chem.* 1976, **251**, 1088.
38. D. A. Julin, H. Wiessinger, M. D. Toney and J. F. Kirch, *Biochemistry* 1989, **28**, 3815.
39. D. A. Julin and J. F. Kirch *Biochemistry.* 1989, **28**, 3825.
40. W. T. Jenkins, *J. Biol. Chem.* 1961, **236**, 1121.
41. H. Batiz-Hernandez and R. A. Bernheim, *Progr.NMR Spectrosc.* 1967, *3*, 63.
42. J. C. Gore and R. P. Kennan, in *Magnetic Resonance Imaging*; Stark, D. D., Bradley, W. G. J., eds; Mosby: St. Louis, 1999; vol. 1, pp 33.
43. E. Tóth, L. Helm and A. E. Merbach, *Top. Curr. Chem.* 2002, **221**, 61.
44. H. Sies, *Metabolic Compartmentation*, H. Sies, Ed., Academic Press, New York, 1982, pp 1.
45. G. Otting, *Prog. NMR Spec.* 1997, **31**, 259.
46. G. Melacini, A.M. Bonvin, M. Goodman, R. Boelens and R. Kaptein, *J. Mol. Biol.* 2000, **300**, 1041.

COMPARISON OF TWO SEQUENCES: SPIN-ECHO AND GRADIENT ECHO FOR THE ASSESSMENT OF DOUGH POROSITY DURING PROVING

A. GRENIER[1,2], T. LUCAS[1], A. DAVENEL[1], G. COLLEWET[1], A. LE BAIL[2]

[1]CEMAGREF, 17 Avenue de Cucillé CS 64427, Rennes, 35044 France
[2]ENITIAA (UMR CNRS 6144), Rue de la Géraudière, BP 8225, Nantes, 44322 France

1 INTRODUCTION

The final character of most bakery products depends to a significant extent on the creation and control of gas bubble structures in the unbaked matrix and the retention of these gas bubbles in a suitable form until the matrix becomes set or baked. Breadmaking may be viewed as a series of aeration stages: bubbles are incorporated during mixing (up to 15% depending on the mixing process and the dough formulation, the bubbles, i.e. the air fraction, are inflated with carbon dioxide produced by the yeast during proving (the end of proving is marked by an expansion up to three or four times its initial volume, and the aerated structure is modified and set by baking.

Proving is a key stage in the development of the final structure of bread. As invasive measurements may provoke dough collapse, the characterisation of the expansion process has been reduced to a number of global volumetric parameters (dough volume, CO_2 volume, etc.). But better understanding and better control of the nucleation and growth of bubbles require the development of non-invasive methods of measurement, with the characterisation of bubble size and distribution. Its feasibility at different scales of observation has been recently investigated with: X-ray tomography[1], confocal microscopy[2] and Magnetic Resonance Imaging (MRI)[3]. Many of these previous works give only qualitative information.

Recently, a MRI method has been presented for estimating the local dough porosity during proving and validated at dough scale by a direct comparison with volumetric measurement[4]. Nevertheless, for long proving times (porosity close to 0.7), the MRI mean porosity deviated from the volumetric mean porosity. The authors suggested that such method could be validated for higher porosities by increasing the signal to noise ratio. This could be done by decreasing the resolution of MR images, increasing the number of average or decreasing the echo time.

The present work aims at evaluating the impact of the echo time on the performance of MRI in quantifying local porosity in dough. The method has already been validated on a spin-echo sequence, and two different echo times (11 and 15 ms) will be tested in the present work. As almost all protons have relaxed within 50 ms for the conditions of the proving process and for the retained formulation[5], very short values of echo time are required to markedly improve the signal to noise ratio. The performance of a gradient echo sequence with shorter echo time (4 and 9 ms) has then also been tested.

2 MATERIALS AND METHODS

2.1 Determining dough porosity with MRI intensity

The method used for estimating the dough porosity from the MRI intensity has already been presented[7].

During proving, the voxel volume is composed of a gas fraction (mainly CO_2) and a paste fraction:

$$V_i(t) = V_{i,gas}(t) + V_{i,pa}(t) \quad (1.)$$

where $V_{i,gas}(t)$ and $V_{i,pa}(t)$ define the volume occupied by gas and paste respectively in the voxel i. The gas volume fraction in the voxel i is called porosity and defined in m^3 of gas at t per m^3 of dough at t (equation 2). By deduction, the paste volume fraction is $1 - \varepsilon_i(t)$.

$$\varepsilon_i(t) = \frac{V_{i,gas}(t)}{V_i(t)} \quad (2.)$$

The grey level of the voxel i composed of both gas and paste can be defined as:

$$GL_i(t) = [1 - \varepsilon_i(t)] GL_{pa,i}(t) \quad (3.)$$

where GL_i was the grey level of voxel i, ε_i its porosity and $GL_{pa,i}$ the grey level of voxel i full of paste.

Subject to the validity of the hypotheses mentioned below:

- the absence of interaction between the gas and the MRI signal of paste is assumed (H1);
- a centred Gaussian distribution of the electronic noise dispersion is assumed (H2);
- the NMR properties of paste are considered to be homogeneous through the whole dough (H3);
- the NMR properties of paste are finally assumed to remain constant with proving time (H4) i.e. temperature or concentration variations or enzymatic reactions do not modify the MRI signal of paste and the grey level of paste is constant during proving;

then the determination of local porosity from the grey level GL_i depends only on the knowledge of GL_i and ε_i at a reference time t_{ref}:

$$\frac{GL_i(t)}{GL_i(t_{ref})} = \frac{1 - \varepsilon_i(t)}{1 - \varepsilon_i(t_{ref})} \quad (4.)$$

The porosity of each voxel i during proving can then be estimated as follows:

$$\varepsilon_i(t) = 1 - [1 - \varepsilon_i(t_{ref})] \frac{GL_i(t)}{GL_i(t_{ref})} \quad (5.)$$

Equation 5 can not be validated easily if ever. In order to validate such method a volume balance could be applied at dough scale. At this scale, the mean porosity $\bar{\varepsilon}(t)$ can easily be estimated by means of volumetric measurements:

$$\bar{\varepsilon}(t) = \frac{V(t) - V(t_{ref})[1 - \varepsilon(t_{ref})]}{V(t)} \quad (6.)$$

where $V(t)$ is the total dough volume and $V(t_{ref})[1 - \varepsilon(t_{ref})]$ is the volume occupied by the dough in the complete absence of gas bubbles. $V(t) - V(t_{ref})[1 - \varepsilon(t_{ref})]$ then represents the volume occupied by gas at time t in agreement with equations 1 and 2. The generalisation of equation 3 to the total dough volume gives:

$$\frac{\overline{GL}(t)}{\overline{GL}(t_{ref})} = \frac{1 - \bar{\varepsilon}(t)}{1 - \bar{\varepsilon}(t_{ref})} \quad (7.)$$

Then, provided that initial mean porosity is assessed, the mean grey level in dough at time t noted $\overline{GL}(t)$ is directly related to the mean porosity at time t. Replacing equation 6 into equation 7 yields to:

$$\frac{\overline{GL}(t)}{GL(t_{ref})} = \frac{V(t_{ref})}{V(t)} \tag{8.}$$

This new relation is of major importance as it will allow the validation of the present method without having to estimate the initial mean porosity. The total dough volume is estimated from the MR image after thresholding:

$$V(t) = n(t) V_i = n(t) \Delta x^2 \Delta z \tag{9.}$$

where V_i is the volume of the voxel i and n(t) the number of voxels assigned to the dough after thresholding, Δx the square pixel side and Δz the slice thickness. Replacing equation 9 into equation 8 implied that the grey level sum should be constant with time[4]. This is a key point for the validation of the method as the grey level sum constancy allows to verify the validity of the assumptions made when developing the method (hypotheses 1 to 5)[4].

2.2 Dough preparation and experimental device

The dough samples were obtained by mixing 500 g of flour, 290 g of water, 12.5 g of salt and 8 g of dry baking yeast in a Kenwood mixer for 4 min at 40 rpm followed by 11 min at 80 rpm. Before incorporation, the yeast was hydrated for 15 min at 35°C in 290 g of water previously mentioned. At the end of mixing, the dough temperature was $24.0 \pm 0.4°C$. Four cylindrical flasks were filled with dough (40.1 ± 0.2 g), closed, weighed and placed in a temperature-controlled environment for proving. Three of them were placed in a temperature controlled laboratory room. Temperature was regulated at $25.6 \pm 0.1°C$. At different times of proving, the total dough volume was measured by reading graduations on the flask. At the same time, the last flask was placed in the measurement area for MRI, the temperature of which was controlled at $26.0 \pm 0.3°C$. At the end of proving, the dough temperature was checked and the flasks were weighed. The weight loss was inferior or equal to 0.3 % of the initial weight. The experiment was repeated twice (3 runs).

2.3 MRI

2.3.1 MRI sequence

The MR images were acquired on a SIEMENS OPEN 0.2T imager and with four sequences: two gradient echo and two spin-echo sequence with an echo time respectively equal to 4, 9, 11 and 15 ms and noted GE4, GE9, SE11 SE15. For all sequences, the following parameters were used: repetition time 300 ms, slice thickness 12 mm, slice position +1.2 mm, FOV size 175 × 175, matrix size 96 × 128 (rectangular FOV 6/8), Gain 89.86, reconstruction scale factor 2.921, bandwidth 279 Hz/pixel, 78Hz/pixel, 130 Hz/pixel respectively for GE4, GE9 and SE. The pixel resolution is equal to 1.37 mm (FOV / matrix size). The FOV was selected to cover the dough height until the end of proving. The average number was equal to 3 (only 2 for the SE11) and the acquisition time was approximately 1 min 30 s (only 1 min 01 s for the SE11).

2.32 Analysis of MR images

To select the volume corresponding to the object under study, the dough (equation 9), the Otsu method was applied[6]. MR images were corrected from nonuniformity by dividing them by a phantom MR image of oil acquired with a TR greater than or equal to 1000 ms and approximately the same echo time[7]. The phantom of cylindrical shape (100 mm in diameter, 120 mm in height) was filled with vegetable oil stabilised at 16°C for one day. The phantom MR images were acquired with each sequence used for the dough MR images. All sequence parameters remained identical (echo time, slice position, FOV, etc.) except TR = 1000 ms and average number = 10. The so-called normalised grey level was calculated from the original grey level value as follows:

$$\forall t \geq 0, \quad GL_{i,n}(t) = GL_{c,p} \frac{GL_i(t)}{GL_{i,p}}$$

(10.)

where $GL_{i,n}(t)$ is the normalised grey level of the voxel i, $GL_{i,p}$ is the grey level of the phantom voxel at the same location as the voxel i in the dough MR image, $GL_{c,p}$ is the grey level of the central voxel in relation to the antenna, which is chosen as a reference (i.e. minimal heterogeneities) and $GL_i(t)$ the grey level of the voxel i in the original MR image.

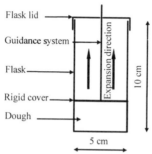

Figure 1   *Experimental device.*

3 RESULTS

If all hypotheses put forward in the method section were satisfied and if the number of voxel assigned to the dough was exact, the mean grey level ratio in dough would then be expected to equal the reciprocal of the total dough volume ratio as presented by equation 8. The grey level ratio is represented as a function of the volume ratio in Figure 2. The reference time was the first MRI acquisition, which occurred 9 min after the end of mixing. The volume considered in Figure 2 is the one calculated from the number of voxels assigned to the dough after the thresholding method has been applied to the original image.

Results from SE11, SE15 and GE4 showed a linear behaviour as expected; the values of the slope was close to unity, respectively equal to 0.998, 0.993 and 0.985 and $R^2 > 0.995$ (the intercept with the y-axis being forced to zero). However, deviation from linearity was observed for long proving times (magnification in Figure 2). Additionally results from GE9 did not present a linear behaviour at all (Figure 2). The objective of the next sections is to identify the sources of such deviations from linearity. Errors in the mean grey level value $\overline{GL}(t) = \sum_{i=1}^{i=n(t)} GL_{i,n}(t) \Big/ n(t)$ can be generated by a bias in the estimation in the number of voxels assigned to the dough and/or of a bias in the grey level sum. The number of voxels assigned to the dough also affects the value of the total dough volume (right term in equation 8).

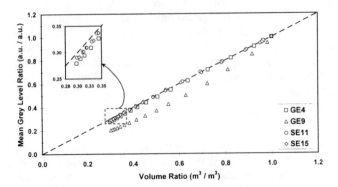

Figure 2 *Evolution of the mean grey level normalised to its initial value in function of the MRI total dough volume ratio estimated after the thresholding method. The dash line represents the first bisector which corresponds to the expected evolution if hypotheses 1 to 5 were satisfied and if the number of voxels assigned to the dough was accurate.*

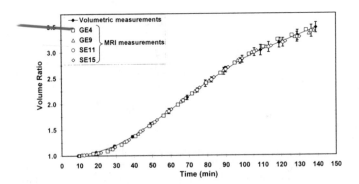

Figure 3 *Comparison between MRI and volumetric measurements: evolution during proving of the total dough volume normalised to its initial volume. The MRI experimental volume represents the mean value of three runs, only the SE11 sequence is represented with its corresponding standard deviation. The volumetric measurement corresponds to the mean value of nine runs (three replicates × three repetitions) with its corresponding standard deviation.*

3.1   Number of voxels assigned to the dough in the MR images

The accuracy of the number of voxels assigned to the dough once the thresholding method had been performed on the original MR images, is evaluated by comparing the total dough volume with the MRI slice volume. It has been previously demonstrated that the two volume ratios $V(t)/V(t_{ref})$ were comparable[4]. The time-course changes in the volume ratio $V(t)/V(t_{ref})$ from MRI and volumetric methods are compared in Figure 3.

For short proving times (t < 30min), the MRI volume ratio was inferior to the total dough volume ratio. This could be attributed to the volume partial effect. For such short times, the voxels of the dough outlines exhibited a grey level value intermediate between the grey level of the background and the mean grey level of dough. As a consequence the threshold value was high compared to the grey level value in the voxels of the dough outlines and the latter voxels were not all taken into account for the dough. This underestimation resulted in an underestimation of the volume at time t and thus of the MRI volume ratio. For longer proving times, both the mean grey level value and the threshold value decreased. As a consequence, the relative number of voxels omitted from the dough outlines was reduced. So the relative error between the two methods (MRI and volumetric measurements) was included in the standard error (< 3.4 %) calculated on the total dough volume ratio from volumetric measurements (Figure 2) and was thus judged non-significant in the present case. So by comparing these two volume ratios and their errors, the segmentation method could be validated and the estimation of the numbers of voxels assigned to the dough at t was judged satisfactory whatever the MRI sequence.

3.2   Constancy of the grey level sum

As the MRI volume had been validated for each sequence, the grey level sum of each sequence was expected to be constant with time during proving (equations 8 and 9). The latter is illustrated in Figure 4. For the SE11 sequence, the grey level sum was constant and close to its initial value during the first 100 minutes. Afterwards, the sum progressively decreased to reach 98 % of its initial value. In the present study, the temperature gradient did not exceed a few degrees (<3°C) and the contribution of temperature variations to the time-course changes in the MRI signal was neglected. Additionally the weight loss was inferior or equal to 0.3 % of the initial weight so its effect could also be neglected. Finally, the enzymatic or metabolic reactions did not interfere with the relaxation[4]. So the decrease of grey level sum for long proving time could be attributed to an underestimation of the number of voxels assigned to the dough. Imperfections in correcting the magnetic inhomogeneities could also be under concern. The same evolution was observed for the SE15 sequence. This suggested that the decrease of the signal to noise ratio due to an increase of the echo time did not affect the grey level sum evolution with the proving time. On the contrary, for the GE4 and GE9, their grey level sum started to decrease at earlier proving times (respectively 40 min and 20 min) to reach at the end of proving 94 and 67% of their initial value. This decrease was attributed to the difference in magnetic susceptibility between paste and gas, which involves the invalidity of hypothesis H1 (there is no interaction between the gas and the MRI signal of paste). Local deformations in the magnetic field (induced by difference in magnetic susceptibility) are all the more intense as gas bubbles get numerous and as the echo time is increased. Both trends can be observed from the present data. In the case of the GE4 sequence, and even for prolonged proving times, such interaction could be neglected as only 6% of the MRI intensity were lost at maximum.

Then the constancy of the grey level sum was validated for the GE4, SE11 and SE15 sequences and for proving times not exceeding 100 min.

3.3   Determination of the mean porosity from the MRI intensity during proving

The comparisons between the mean porosities calculated from MRI intensity (equation 7) and from volumetric measurements (equation 5) are now presented as a function of proving time in Figure 5. The reference was still the first MRI acquisition. As gas production and/or diffusion is commonly retarded up to 20 min after the end of mixing, porosity at t_{ref}, $\bar{\varepsilon}(t_{ref})$, was assumed not to have changed from its initial value (at the end of mixing, t=0) and was set at 10% (m^3 of gas per 100 m^3 of dough). MRI mean porosity was calculated from GE9 results in order to illustrate the extent of deviation that can be generated if hypothesis H1 was not satisfied. For proving time superior to 30 min, the porosity estimated with the GE9 was overestimated (by 8 to 12% with the reference to the mean porosity calculated from volumetric measurements). These results confirmed the non-suitability of such MRI sequence to monitor the porosity in dough during proving. They also focused on the need for potential users to control the validity of hypotheses H1 to H5 before applying the MRI method.

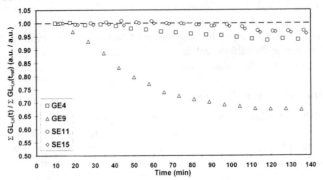

Figure 4   *Evolution of the grey level sum normalised by its initial value. The dash line represents the unity which corresponds to the expected constancy.*

Mean porosities calculated from GE4, SE11 and SE15 results exhibited a general trend very similar to the mean porosity calculated from volumetric measurements. For short proving times, whatever the sequence, the MRI porosity was inferior to the "volumetric" porosity. The relative error between the two methods of calculation was over 5% with a maximum of 16 and 9% respectively for GE4 and SE11. For the SE15 sequence, the relative error was inferior to 5%. Whatever the sequence this could be attributed to an error on the mean grey level observed at the reference time which could be induced by an error on the number of voxels[1]. For proving times over 30 minutes, the MRI porosity was slightly superior to the "volumetric" porosity. But the relative error between the two methods of calculation was under 5 %. At the end of proving (t > 100 min), slightly better estimations were obtained from the SE sequences (overestimation close to 2 % expressed in m^3 of gas per 100 m^3 of dough) than from the GE4 sequence (overestimation close to 3%). So the non constancy of the grey level sum observed for long proving time (Figure 4) may induce a relative error of 4%. So better results were obtained with SE11 and SE15 than with GE4 although the signal to noise ratio was better for GE4 (signal to noise ratio respectively in the range [11, 42], [10, 37.5] and [9.9, 37] for GE4, SE11 and SE15).

Decreasing the echo time for the SE sequence did not improve the accuracy of the mean porosity. This can be explained by an equivalent signal to noise ratio (cited above). In fact, the signal intensity was higher for the SE11 but its noise level was also higher due to a lower average number (2 versus 3). Using an average number of 3 for the SE11 should have increased the signal to noise ratio by a factor of 22% ($\sqrt{3/2}$). No real conclusion on the effect of the echo time for a SE sequence can be drawn, but it could be expected that using an average number of 2 for the SE15 would have degraded the results.

Decreasing the echo time when passing from the SE sequence to the GE sequence degraded the accuracy of the mean porosity. But this phenomenon was attributed to the sequence itself which was sensitive to the interaction between the gas and the MRI signal of paste. The effect of decreasing the echo time from 11 to 4 ms could not be evaluated.

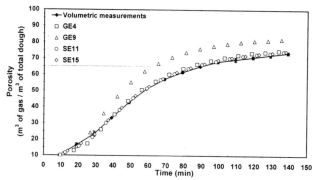

Figure 5 *Comparison between MRI and volumetric measurements: evolution of the mean porosity in dough during proving. The volumetric porosity is equal to the mean value of three runs with its corresponding standard. The MRI porosity of each sequence is estimated directly from the grey level ratio (equation 7).*

4 CONCLUSION

The present work aimed at evaluating the performance of two MRI sequences: spin-echo and gradient echo sequences with different echo times, to quantify local porosity in dough. Results suggested that a gradient echo must be used very carefully when attempting to assess the porosity in dough and recommended the use of very short echo times ($\leq 4ms$). Quantifying the effect of the echo time passes through the development of a spin echo sequence with shorter echo times ($\leq 11ms$) –with fixed average number, bandwidth, etc. As the aim of the present method is to quantify the local porosity in dough, the different contributions to its uncertainty should be also estimated in future studies, in particular the varying contribution of the signal to noise ratio.

References

1. M.B. Whitworth and J.M. Alava, *Bubbles in Food*, 1st edn, K. Eds. Eagan Press, St. Paul, 1999, Section 3, p 221.
2. K.H. Sutton, L.D. Simmons, M.P. Morgenstern, A. Chen and T.L. Crocker, 51st RACI Cereal Chemistry Conference, Sydney, September 2001.
3. H. Takano, N. Ishida, M. Kiozumi and H. Kano, J. Foof Science, 2002, Vol 67, **1**, 244.
4. A. Grenier, T. Lucas, G. Collewet., A. Lebail, "Assessment by MRI of local porosity in dough during proving. Theoretical considerations and Experimental validation using a Spin-Echo sequence.", submitted to Magnetic Resonance Imaging.
5. A. Grenier, T. Lucas, A. Davenel, A. Lebail, M. Cambert, "NMR assessment of relaxation time and ice fraction in frozen dough", submitted to Journal of Cereal Science.
6. N. Otsu, IEEE, 1979, Vol. SMC-9, **1**, 62.
7. G. Collewet, A. Davenel, C. Toussaint and S. Akoka, MRI, 2002, Vol 20, **4**, 365.

SOLID-STATE NMR OF LYOTROPIC FOOD SYSTEMS

E. Hughes, P. Frossard, L. Sagalowicz, C. Appolonia Nouzille, A. Raemy, H. Watzke

Nestlé Research Centre, Nestec Ltd., Vers-chez-les-Blanc, CH-1000 Lausanne 26, Switzerland

1 INTRODUCTION

Lyotropic materials possess certain properties that make them attractive as building blocks within new food products[1]. For example, they possess a large surface area with both hydrophilic and hydrophobic domains, a three-dimensional structure of various length scales (nanometer to micrometer) and phases that form spontaneously and are stable in excess water.

In this article, we demonstrate the utility of solid-state NMR to study lyotropic materials formed from food grade monoglyceride mixtures. NMR techniques that have historically been applied to pure component lyotropic systems are used to study the food grade equivalent. In the first instance, the use of carbon-13 chemical shift anisotropy[2] to distinguish the phase after the addition of guest molecules is discussed. Finally, the utility of double-quantum[3-5] deuterium NMR is demonstrated in regions of the phase diagram where both anisotropic and isotropic phases are present.

2 MATERIALS AND METHODS

The food grade monoglyceride mixture, Dimodan$_{LS}$, (Danisco, Copenhagen Denmark) was used in the studies. The major components of Dimodan$_{LS}$ are given in Table I. Samples for use in the NMR studies were prepared in 5.0 g batches. The monoglyceride was melted in a beaker and then the appropriate amount weighed into a 20 mL vial with a screw cap top. Deuterated water (99.8 % Aldrich) was then added. A magic angle sample spinning rotor, either 4 or 7 mm was then placed in the vial and the top screwed on tightly. The sample was repeatedly heated with an air gun to above 100 °C and mixed using a table top vortexer. The sample was allowed to cool to room temperature and left to settle for more than a day. Afterwards, the rotor was extracted from the vial, the outside cleaned and the spinning cap pressed onto it. Before and after the NMR experiments the samples were routinely weighed to see if the weight of the sample remained constant.

All NMR experiments were performed on a Bruker DSX 400 (^{1}H Larmor frequency 400.13 MHz). Depending on availability, experiments were carried out using either a 7 mm double resonance CP-MAS probe or a triple resonance 4 mm CP-MAS probe. All experiments were performed with the magnet unlocked. For the deuterium experiments, typical 90° pulse widths were around 7 μs with recycle delays set to 2 s for single pulse experiments and 5 to 10 s for the double quantum experiments.

Differential scanning calorimetry measurements (DSC) were performed on a Micro-DSC III (Setaram, Caluire France) operated in scanning mode. The reference material was Al_2O_3. Two heating scans were performed over the temperature range of 10°C to 110°C. The heating rate was 0.2°C min⁻¹. The sample weight was approximately 200 mg.

Table I *Composition of Dimodan$_{LS}$*

Fatty acid carbons:double bonds			% composition	
saturated {	16:0		6.0	} 10.6 %
	18:0		4.6	
unsaturated {	18:1	1-monoolein	21.3	} 87.4 %
	18:2	1-monolinolein	66.1	
	Others		1.4	

3 RESULTS AND DISCUSSION

3.1 Dimodan$_{LS}$ and Water Phase Diagram

The two major components of Dimodan$_{LS}$ (D$_{LS}$) are both unsaturated, 1-monolinolein at approximately 66 % and 1-monoolein at 21 %. The double bonds occur at carbon positions nine and twelve and position nine respectively. When water is added to the Dimodan$_{LS}$ various phases are formed depending on composition and temperature. An idealised phase diagram is shown in Figure 1 derived from polarised light microscopy work[1]. A number of phases were identified which included lamellar L$_\alpha$, inverse micellar L$_2$, inverse hexagonal H$_{II}$ and cubic. At high concentrations of water, a two phase system exists. Depending on temperature, this phase consists of L$_2$, H$_{II}$ or cubic with free water.

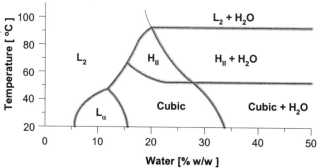

Figure 1 *Idealised phase diagram of Dimodan$_{LS}$ and water. The broad lines at the phase boundary represent a mixture of phases.*

3.2   High resolution magic angle spinning and phase detection

High-resolution proton and carbon-13 NMR spectra of the Dimodan$_{LS}$:H$_2$O phases can be acquired by spinning the sample at the magic angle[6]. Typically, a spinning speed of approximately 4 kHz is required for both proton and carbon-13 spectra.

In anisotropic regions of the phase diagram, cross-polarisation (CP) techniques can be used to obtain carbon-13 spectra. In isotropic regions, CP methods do not work due to the influence of isotropic motion that averages the heteronuclear dipolar interactions more efficiently. Therefore, cross-polarisation can be used to distinguish the nature of the phase.

If slow magic angle sample spinning is used, around 400 to 1000 Hz, spinning side bands on the carbonyl resonance are observed in anisotropic regions of the phase diagram. Carbonyl carbon atoms typically have a large chemical shift anisotropy (CSA). The restricted motion in the anisotropic regions is not sufficient to average completely the CSA. Due to the different types of motion present in hexagonal and lamellar phases, the residual CSA is different. Typically the size of the residual anisotropy in the lamellar phase is double that in the hexagonal phase. Furthermore, the sense of the anisotropy is different in the two phases. In principle the spinning sideband pattern contains information on the dynamics of the monoglycerides, but due to the low signal to noise ratio of the spectra the analysis is difficult.

This technique is useful in distinguishing the phase of a system when a guest molecule has been added to the lyotropic system. For example, when sufficient glucose is added to a 80:20 D$_{LS}$:D$_2$0 sample, the phase of the sample changes from isotropic (cubic) to an anisotropic phase (unknown). Identification of the phase from deuterium NMR cannot be done with a single spectrum (Figure 2.c). Polarised light microscopy is also ambiguous (results not shown). Using slow magic angle carbon-13 NMR, focussing on the carbonyl resonance, one can distinguish quite clearly between H$_{II}$ and L$_\alpha$ phases. (Figure 2.a and 2.b). The carbon-13 spectra in figure 2.a and 2.b are from an undoped 80:20 D$_{LS}$:H$_2$O sample at 5 °C and 70 °C. The temperatures correspond to L$_\alpha$ and H$_{II}$ anisotropic regions respectively. One can see that the anisotropy is different in the two phases and that the spinning sidebands in the L$_\alpha$ phase cover twice the spectral width as the H$_{II}$. The carbonyl carbon-13 spectrum in figure 2.d corresponds to the doped glucose sample at a level of 10 % loading, from the sense of the anisotropy and the span of the spinning side bands one can quite clearly see that the phase is hexagonal.

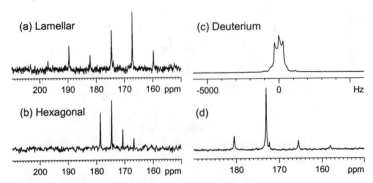

Figure 2   *The use of carbonyl ^{13}C CSA to distinguish the phase. (a) and (b) undoped 80:20 D$_{LS}$:H$_2$0 at 70 and 5 °C. (c) ^{2}H, and (d) ^{13}C, doped glucose sample of 80:20 D$_{LS}$:D$_2$0 at 25 °C.*

3.3 DSC results of Dimodan$_{LS}$:D$_2$O 80:20 sample

The lyotropic phases obtained from D$_{LS}$ and water are sometimes unstable. The application of a shear force or shock can induce crystallization in the cubic phase. Also aged samples will show signs of crystallization. This was clearly evident in freshly prepared samples used for DSC analysis. Just putting the sample in the DSC instrument can induce phase changes. Figure 3 shows two consecutive heating runs on a D$_{LS}$:D$_2$O sample of mass ratio 80:20. In the initial heating run there is a large endothermic peak around 33 °C with an enthalpy of 14.9 J g^{-1}. Also one can see endothermic transitions for the two expected phase transitions of cubic to H$_{II}$ and H$_{II}$ to L$_2$ at 60 and 92 °C with enthalpies of 0.29 J g^{-1} and 0.54 J g^{-1} respectively. In the second heating run the endothermic transition around 33 °C is significantly diminished with only the two expected phase transitions observed with comparable enthalpies to the first run.

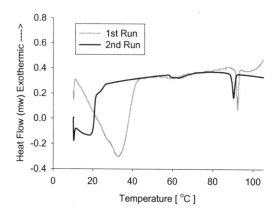

Figure 3   *DSC experiments on 80:20 D$_{LS}$:D$_2$O. The first run shows an endothermic peak at 33 °C which is not present in the second run. The expected phase transitions at 60 and 90 °C are observed and are reversible.*

Normally static single pulse deuterium NMR is very good at following the phase change in such lyotropic systems when deuterated water is used[7]. In the anisotropic region the lineshape is averaged considerably so the anisotropic linewidth of the signal is on the order of 2 to 5 kHz therefore the signal to noise ratio of the experiment is high. Quality data can be obtained in very few scans. However, it was very difficult to observe by NMR, the anisotropic crystals present in the cubic phase that were evident from DSC and polarised-light microscopy (data not shown). Two reasons for this were that first, the preparation method for the NMR experiments produced an undisturbed sample free from shear leading to a "clean" cubic phase. Second, the signal from any anisotropic material present would be small compared to the main isotropic cubic signal. In order to reveal the anisotropic character of the phase we have used double quantum deuterium NMR.

3.4 Double quantum deuterium NMR

In a series of recent papers Navon and co-workers[8-10] have shown that deuterium double quantum NMR can be used to differentiate deuterated water in anisotropic and isotropic environments. Their studies on ordered biological tissues clearly demonstrated the removal of the large isotropic water signal to reveal anisotropic water. They were able to distinguish various water signals from different anisotropic environments. Applying their methodology we have been able to successfully isolate the anisotropic signal from the dominant cubic isotropic signal.

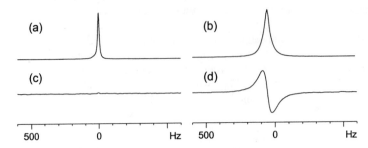

Figure 4   *Comparison of single (a,b) and double quantum (c,d) deuterium NMR experiments for two different experimental conditions for a single sample of 80:20 D_{LS}:D_2O. All spectra were recorded at a temperature of 25 °C. Spectra (a,c) were recorded on the fresh sample. Spectra (b,d) were recorded on a sample that had been first cooled to −20 °C and slowly warmed up to room temperature. The DQ experiments (c,d) are on the same vertical scale.*

Figure 4 shows the results from two experiments. In the first experiment, the single (4.a) and double (4.c) quantum deuterium NMR spectra were recorded on a freshly prepared 80:20 D_{LS}:D_2O sample at 25 °C. The single quantum NMR spectrum shows a narrow Lorentzian resonance characteristic of an isotropic phase. The double quantum spectrum (4.c) has only a very small signal showing that the phase is isotropic. The residual signal is due to incomplete removal of the isotropic signal. Spectra (4.b) and (4.d) are from the same sample and at the same temperature as the previous experiments. They again show the single (4.b) and double (4.d) quantum deuterium spectrum. However, before the spectra were recorded the temperature of the sample was lowered to −20 °C and raised slowly back to room temperature. The single quantum spectrum (4.b) shows again a Lorentzian lineshape but is wider than spectrum (4.a), 55 Hz compared to 15 Hz. This change in line width could be accounted for in a number of ways. However, the double quantum spectrum clearly shows that it is due to an anisotropic component within the sample.

3.4 ^{2}H Double quantum NMR in the cubic + water region.

The double quantum NMR technique can also be used in regions of the phase diagram where free water is present in equilibrium with a second phase of the dimodan$_{LS}$. A sample was prepared with a mass ratio of 50:50 D_{LS} to D_2O. From the phase diagram obtained by polarized light microscopy, the sample should be composed of a cubic phase plus free water at room temperature. When this sample was prepared for the NMR experiments shearing occurred in the sample as it was put into the NMR rotor. Therefore, one would expect the phase may consist of both isotropic and anisotropic phases plus pure water. The single quantum spectra (not shown) consisted of single isotropic line. However a double-quantum signal could be produced as shown in Figure 5.

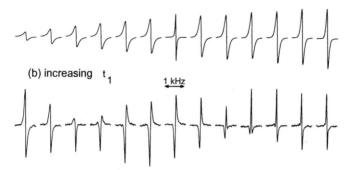

Figure 5   *Double quantum signal of $D_{LS}:D_2O$ (50:50); (a) optimisation of DQ*
preparation period. (b) evolution of DQ signal when transmitter offset by 100
Hz. Double quantum pulse sequence used $90-\tau/2-180-\tau/2-90-t_1-90-Acq$[7].

Figure 5.a shows the optimisation of the DQ build-up for the sample at room temperature.
While Figure 5.b shows the oscillation of the DQ signal when the transmitter is placed
100 Hz off resonance. This oscillation can be fit to the following equation[8] to give the
oscillation frequency and the double quantum relaxation time.

$$I = I_0 \cos(4\pi\Delta v_0 t_1)\exp(-t_1/T_{DQ}) \tag{1}$$

The results of the fit are shown in Figure 6. The Δv_0 was found to be 97 Hz and the double
quantum relaxation time T_{DQ} to be 11 ms for a τ value of 4 ms. This data confirms the
double quantum nature of the signal and that it is not due to breakthrough from the large

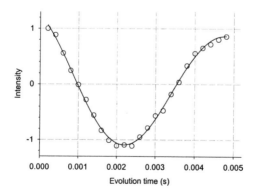

Figure 6   *Oscillation of the signal intensity as a function of DQ evolution time, t_1, for*
an offset frequency of 100 Hz.

isotropic water signal. Although the single quantum signal gave what would appear to be an isotropic line, it is clear from the double quantum experiments that the phase is not solely isotropic.

4 CONCLUSIONS

In this paper we have tried to demonstrate that even with lyotropic materials where the surfactant material is a mixture, one is able to obtain useful information on the phase of the systems and follow the phase behaviour as a function of temperature, composition and time. The double quantum NMR experiments can be used to observe anisotropic phases forming in isotropic cubic phases or where a dominant isotropic water signal exists. Also, one should be able to extract dynamic parameters on the anisotropic water component[11]. The carbon-13 NMR studies also show that the phase of the system can be monitored and identified and could be used in studies where the use of deuterated water is inconvenient.

References

1.   S. Vauthey, *PhD Thesis*, 1998.
2   B. A. Cornell, *Chem. Phys. Lipids*, 1981, **28**, 69.
3   G. Bodenhausen, *Prog. NMR Spectrosc.* 1981, **14**, 137.
4   I. Furo and B. Halle, *Chem. Phys. Lett.* 1991, **182**, 6.
5   G. Jaccard, S. Wimperis and G. Bodenhausen, *J. Chem. Phys.*, 1986, **85**, 6282.
6.   A. Pampel, E. Strandberg, G. Lindblom and F. Volke, *Chem. Phys. Lett.* 1998, **287**, 468.
7   W. G. Morley and G J. T. Tiddy, *J. Chem. Soc. Faraday Trans.*, 1993, **89**, 2823.
8   H. Shinar, Y. Seo and G. Navon, *J. Magn. Reson.* 1997, **129**, 98.
9   U. Eliav and G. Navon, *J. Magn. Reson. A*, 1995, **115**, 241.
10   Y. Sharf, U. Eliav, H. Shinar and G. Navon, *J. Magn. Reson. B*, 1995, **107**, 60.
11   C. Y. Cheng and L. P. Hwang, *J. Chin. Chem. Soc.*, 2001, **48**, 953.

APPLICATION OF NMR AND HYPHENATED NMR SPECTROSCOPY FOR THE STUDY OF BEER COMPONENTS

I. F. Duarte[1], M. Spraul[2], M. Godejohann[2], U. Braumann[2] and A. M. Gil[1]

[1] Department of Chemistry, University of Aveiro, 3810-193 Aveiro, Portugal
[2] Bruker BioSpin GmbH, Silberstreifen, D76287 Rheinstetten, Germany

1 INTRODUCTION

Beer is a fermented beverage made from malted grains (usually barley), hops, yeast and water, and contains a vast number of compounds widely ranging in nature and concentration level[1]. There has been great interest in studying the chemical composition of beer, as this information is essential for quality assessment and for the development of new products. Most data reported in the literature has been obtained with basis on analytical techniques that often involve some kind of pre-treatment of the beer sample in order to concentrate the desired group of compounds, making the process of overall characterisation of beer very laborious and time consuming. High resolution NMR has already proven useful for studying the composition of different liquid foods such as fruit juices[2-4], coffee[5] and wine[6-8], offering the advantages of giving information about a very wide range of different components in a single experiment, in which the sample is analysed directly and non-invasively.

A recent NMR application to beer has revealed the specificity of the technique towards beer composition[9]. In addition to the application of high resolution NMR, further investigation of some beer components, namely carbohydrates and aromatic compounds, may be carried out by HPLC-NMR/MS, as described in the present paper. Carbohydrates are the major non-volatile components of beer, comprising mainly fermentable sugars (e.g. glucose, maltose, maltotriose) and dextrins with varying degrees of polymerisation (DP) and branching patterns. They have a marked influence on beer taste and body, and thus, the development of rapid and efficient methods for monitoring the beer carbohydrate composition is of paramount importance for modern brewing technology. The aromatic composition of beer is also investigated, in order to overcome the assignment difficulties found when using NMR alone, related to the low abundance of these substances and the high degree of overlap in the aromatic region of the ^{1}H spectrum. HPLC-NMR/MS combines the separation ability of chromatography with the powerful structure determination ability of NMR and mass spectrometry, thus having a great potential for the characterisation of complex mixtures, as has been extensively demonstrated in the area of pharmaceutical research[10]. In the area of food analysis, however, only a few applications have been reported. These include an example of on-flow HPLC-NMR of a wine concentrate[11], the investigation of the detailed structures of hop and beer bitter acids by

HPLC-NMR[12], and the identification of several quercetin and phloretin glycosides in an apple peel extract by HPLC-NMR/MS[13].

2 MATERIALS AND METHODS

2.1. Sample Preparation

For NMR analysis, the beer sample (ale type) was degassed in an ultrasonic bath (10 minutes) and prepared to contain 10% D_2O (deuterium lock) and 0.02% sodium 3-(trimethylsilyl)-propionate (TSP) as chemical shift reference. For HPLC-NMR/MS analysis of carbohydrates, the beer was simply degassed before injection, whereas for the characterisation of aromatic compounds the sample was concentrated to approximately 2/3 of the initial volume by rotor evaporation at 40°C during 20-30 min.

2.2. NMR Spectroscopy and HPLC-NMR/MS

1D and 2D NMR spectra of beer were recorded on a Bruker Avance DRX-500 spectrometer, operating at 500.13 MHz for proton and 125.77 MHz for carbon. The ¹H 1D spectra were acquired using a pulse sequence based on the two-dimensional NOE experiment, with a 90° pulse of 8.5 µs. Water (4.77 ppm) and ethanol signals (1.17 and 3.64 ppm) were suppressed by applying a shaped pulse, with triple offset and amplitude scaling, during relaxation delay and mixing time (100 ms). 128 transients were collected into 16k data points with a spectral width of 5482 Hz. Total correlation (TOCSY), ^1H-^{13}C correlation and *J*-resolved spectra were also recorded to aid spectral assignment.

For the study of carbohydrates by HPLC-NMR/MS, chromatographic separation was carried out using a cation-exchange ION-300 column, at a flow rate of 0.3 ml/min. The injection volume was 100 µl and the mobile phase consisted of 0.0085N H_2SO_4 in D_2O. ¹H NMR spectra were recorded at 500 MHz using the on-flow and the loop-sampling methods. Electrospray ionization (ESI) was carried out in positive-ionisation mode, with addition of 20 mM aqueous sodium acetate solution to facilitate ionisation.

For the HPLC-NMR/MS study of aromatic compounds, chromatographic separation was carried out using a Purospher reverse phase (RP18) column, at a flow rate of 1.00 ml/min, with diode array detection in the UV-Vis region. The injection volume was 500 µl and the mobile phase consisted of a mixture of D_2O (containing 0.06% formic acid) and acetonitrile, changing in composition according to the following gradient: 0 min - 3% acetonitrile, 40 min - 40 % acetonitrile, 60 min - 100 % acetonitrile, 70 min - 3% acetonitrile. ¹H NMR spectra were obtained at 500 MHz using the time-sliced stop-flow and the loop-sampling methods. ESI-MS analysis of loops was carried out in positive and negative ionisation modes.

3 RESULTS AND DISCUSSION

3.1 Characterisation of Beer Composition by High Resolution ¹H NMR

Figure 1 shows the TOCSY spectrum of a beer sample, as well as the projected 1D spectrum, recorded at 500 MHz with suppression of ethanol and water signals. The general high spectral complexity clearly shows the potential of the technique to enable the identification of many different compounds, present in a wide range of concentrations.

Some of the compounds identified by both 1D and 2D NMR are indicated and a more complete list is given in a recent publication[9].

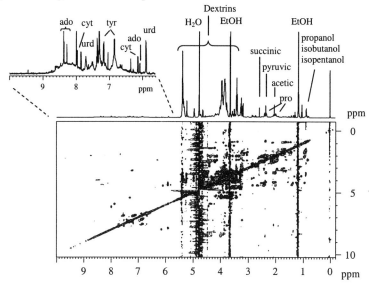

Figure 1 *500 MHz [1]H 1D and TOCSY spectra of beer   ado, adenosine; urd: uridine; cyt: cytosine; tyr: tyrosine; pro: proline*

The high-field region (0.0-3.0 ppm) shows signals arising mainly from alcohols (propanol, isobutanol, isopentanol), aliphatic organic acids (citric, malic, lactic, pyruvic, acetic, succinic) and amino acids (alanine, γ-aminobutyric acid, proline).

The 3.0-6.0 ppm region of the spectrum reflects the strong contribution of beer carbohydrates. This region shows relatively lower resolution, compared to the remaining parts of the spectrum, reflecting the predominance of medium- and/or high-molecular weight dextrins undergoing slow molecular tumbling and, hence, characterised by faster transverse relaxation and broader signals. However, spectral distinction between dextrins of different sizes is extremely difficult because of strong peak overlap in both uni- and bi-dimensional NMR spectra (Figure 1). The spectral similarity for different glucose oligomers is illustrated by the set of 1D ^{1}H NMR spectra of standard solutions shown in Figure 2. The linear carbohydrates maltose, maltotriose and maltoheptaose, consisting of glucose monomers linked by α(1-4) linkages, show similar overlapping profiles. In the anomeric region, the protons H1 of the reducing α- and β-glucose units give rise to the doublets at 5.22 and 4.63 ppm, respectively. The non-reducing H1 protons engaged in α(1-4) linkages resonate at about 5.38 ppm, whereas those involved in α(1-6) linkages are shifted upfield to about 4.96 ppm, as seen for isomaltose (Figure 2b). In the beer spectrum (Figure 2e), the group of overlapped signals in the 5.3-5.4 ppm range reveals the presence of different-sized linear glucose segments, while the 4.96 ppm peak is indicative of branched carbohydrates containing the α(1-6) linkage. However, in such a complex mixture, further assignment is hindered by the broad and overlapped nature of the spectrum, thus calling for improved analytical methods.

The signals situated in the low-field region are the weakest in the spectrum and show a high degree of overlap. With support of 2D spectra, compounds like tyrosine, phenylalanine, tryptophane, uridine (urd), adenosine (ado) and cytosine (cyt) could be identified. Moreover, the underlying broad humps between 6.7 and 8.7 ppm (see insert in 1D spectrum in Figure 1) have been tentatively attributed to polyphenols, which may be associated with non-aromatic moieties, as suggested by some TOCSY correlations with signals in the 3.8-4.5 ppm range.

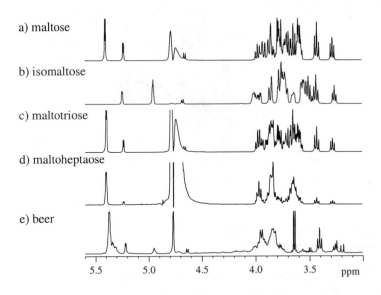

Figure 2 *500 MHz ^1H NMR spectra of standard carbohydrates differing in size and structure (a-d) and of the beer sample (e)*

3.2. Identification of Beer Carbohydrates by HPLC-NMR/MS

The on-flow NMR chromatogram obtained by HPLC-NMR/MS analysis of beer is shown in Figure 3a. Some components are readily separated and identified through their ^1H NMR spectra, namely ethanol (RT 45.0 min), glycerol (RT 30.0 min), fructose (RT 22.5 min) and citric acid (RT 19.5 min). In the case of glycerol, it is noted that its detection by NMR of beer is usually hindered by the fact that its signals (δ 3.48, 3.57, 3.70 ppm) are masked by the carbohydrate region and affected by the ethanol suppression. The major beer carbohydrates elute between 12 and 17 min RT and Figure 3b shows some of the rows extracted from the on-flow record in that range of retention times. The observed profiles reflect the elution of malto-oligosaccharides of different sizes (dextrins), the larger compounds being eluted first, as shown by the poorer resolution of the earlier spectra.

The MS data recorded for the same fractions in positive-ionisation mode (not shown) enable information on the dextrins size to be obtained. The fraction eluting at 12.3 revealed the presence of oligomers with degree of polymerization (DP) of 6, 7, 8 and 9; the signal seen at 4.96 ppm in the corresponding NMR spectrum (Figure 3b) reveals the presence of α(1-6) branched dextrins. The 12.8 min fraction was seen to contain DP5 and DP6 dextrins and, again, the corresponding NMR row indicates ramification of these carbohydrates to some extent (Figure 3b). The fractions collected at 13.7 min and at 14.7 min were found to contain, respectively, maltotetraose and maltotriose. Finally, a subsequent fraction obtained at 16.6 min (not shown) gave an MS peak characteristic of a glucose disaccharide, enabling the identification of maltose and trehalose.

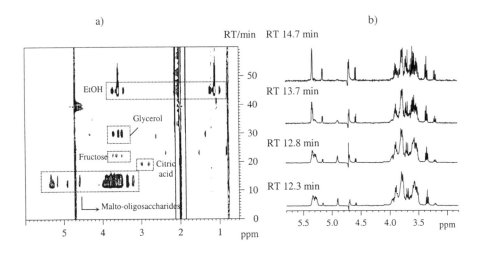

Figure 3 *a) On-flow NMR chromatogram resulting from the elution of beer on the column ION-300 (flow rate 0.3 ml/min; injection volume 100 µl); b) rows extracted from the on-flow record*

3.3. Identification of Beer Aromatic Compounds by HPLC-NMR/MS

The HPLC-NMR/MS analysis of beer aromatic compounds was carried out using the time-sliced stop-flow mode, in which the flow is stopped at short intervals to acquire NMR spectra, and started again after the NMR measurement is finished. By combining the 1D NMR spectra acquired, a "pseudo" on-flow diagram is obtained, allowing an easy visualisation of the number and type of compounds detected. This diagram is shown in Figure 4 for the beer sample studied and reveals the separation of many aromatic components, with signals in the 5.5-10.0 ppm range. Some fractions were analysed by MS after collection into loops during another chromatographic run. With basis on the structural information provided by NMR and by MS, several beer components could be unambiguously identified, as indicated in Figure 4. Besides confirming the presence of many compounds previously identified by 1D and 2D NMR, the method enabled additional compounds to be found e.g. tyrosol, 2-phenylethanol and an unknown *o*-disubstituted alcohol. These compounds had not been identified before, due to strong peak overlapping in the aromatic region and to the inexistence of TOCSY correlations between aromatic and aliphatic side chain protons.

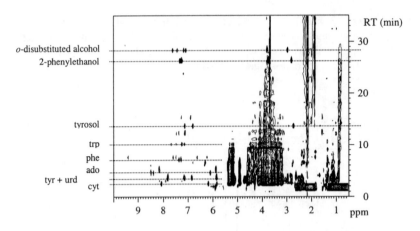

Figure 4 *"Pseudo" on-flow NMR chromatogram, resulting from the elution of beer on the column RP-18 (flow rate 1.00 ml/min; injection volume 500 µl)*

4 CONCLUSIONS

This work has shown that high resolution NMR is of great utility to characterise the complex chemical composition of beer, enabling the rapid identification of a large number of compounds, present in a range of concentrations. This has been achieved in a non-invasive manner, requiring only the degassing of the beer sample. However, spectral distinction of beer carbohydrates constituted of glucose monomers is hindered by peak overlap even in the 2D spectra and a deeper insight may be achieved by HPLC-NMR/MS. This method enabled the identification of several carbohydrates, from small fermentable sugars to dextrins with up to 9 glucose monomers. HPLC-NMR/MS has also been useful for confirming the identity of some aromatic compounds previously assigned by NMR alone and for revealing new ones.

References

1   P.S. Hughes and E.D. Baxter, in *Beer Quality, Safety and Nutritional Aspects*, RSC, Cambridge, 2001.
2   P.S. Belton, I. Delgadillo, A.M. Gil, P. Roma, F. Casuscelli, I.J. Colquhoun, M.J. Dennis and M. Spraul, *Magn. Reson. Chem.*, 1997, **35**, S52.
3   G. Le Gall, M. Puaud and I.J. Colquhoun, *J. Agric. Food Chem.*, 2001, **49**, 580.
4   A.M. Gil, I.F. Duarte, I. Delgadillo, I.J. Colquhoun, F. Casuscelli, E. Humpfer and M. Spraul, *J. Agric. Food Chem.*, 2000, **48**, 1524.
5   M. Bosco, R. Toffanin, D. Palo, L. Zatti and A. Segre, *J. Sci. Food Agric.*, 1999, **79**, 869.
6   A. Ramos and H. Santos, in *Annual Reports on NMR Spectroscopy*, ed. G. Webb, Academic Press, London, 1999, vol. 37, p.179.
7   I.J. Kosir and J. Kidric, *J. Agric. Food Chem.*, 2001, **49**, 50.

8   M.A. Brescia, V. Caldarola, A. De Giglio, D. Benedetti, F.P. Fanizzi and A. Sacco. *Analytica Chimica Acta*, 2002, **458**, 177.

9   I. F. Duarte, A. Barros, P. S. Belton, R. Righelato, M. Spraul, E. Humpfer and A. M. Gil, *J. Agric. Food Chem.*, 2002, **50**, 2475.

10 J.C. Lindon, J.K. Nicholson, I.D. Wilson, *J. Chromatogr. B*, 2000, **748**, 233.

11 M. Spraul and M. Hofmann, in *Magnetic Resonance in Food Science*, eds. P.S. Belton, I. Delgadillo, A.M. Gil and G.A. Webb, Royal Society of Chemistry, Cambridge, 1995, p.77.

12 K. Pusecker, K. Albert, E. Bayer, *J. Chromatogr. A*, 1999, **836**, 245.

13 A. Lommen, M. Godejohann, D.P. Venema, P.C.H. Hollman and M. Spraul, *Anal. Chem.*, 2000, **72**, 1793.

Food: Structure and Dynamics

ASSESSMENT OF MEAT QUALITY BY NMR

J.P. Renou, G. Bielicki, J.M. Bonny, J.P. Donnat and L. Foucat

Structures Tissulaires et Interactions Moléculaire INRA Theix Saint Genès Champanelle 63122France

1 INTRODUCTION

Meat has always been a very important food in developed countries, and the tenderness, moist texture, rich flavour and high nutritional value of muscle make it a highly desirable foodstuff. However, in recent years, new expectations of meat quality have appeared and are still evolving. In industrial societies purchasers expect an optimal quality/price ratio and consistency in quality. This is true for both the processors who buy the raw materials, and for the end consumers who purchase the finished products. This trend has led to increasing demands, particularly in the last three decades, for:

- Technological quality, with the intense industrialisation of meat processing.
- Guarantees of safety and eating quality of food commodities, as consumer choice broadens.
- Authenticity, a concept that combines many factors such as pureness, proper description of the product, and designation of origin.

1.1 What is meat ?

The main constituents of mammalian skeletal muscle and their proportions are: water 65-80%, protein 16-22%, carbohydrate 1-2%, fat 1-13 % and other soluble material 1%. Skeletal muscle is a highly organised (Figure 1) and is made up of groups of muscle fibres enclosed and supported by connective tissue.

The epimysium is the heavy sheath enclosing the entire muscle. Fibre bundles are surrounded by the perimysium, while the endomysium envelops each individual muscle fibre. The muscle fibre is long (1-40 mm) and cylindrical (10-100 μm diameter). Bundles of muscle fibres may run parallel to the main muscle axis or have a certain pennation angle between muscle and fibre directions.

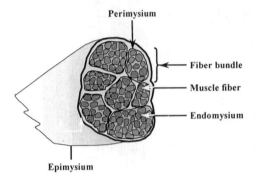

Figure 1   *Diagram of muscle showing the different connective tissues, epimysium, perimysium and endomysium, in relation with the muscle fibres (from Etherington et al.[1]).*

2 TECHNOLOGICAL QUALITY

2.1 WHC

The interactions between water and macromolecules determine the Water Holding Capacity (WHC) of meat. Meat WHC depends primarily on the extent of *post mortem* myofibrillar shrinkage and the correlative changes in the extracellular water compartments.[2,3] WHC of fresh meat was assessed by NMR relaxation measurement of water protons, which yields information on the water dynamics.[4,5] The general features of proton relaxation in muscle are characterised by a longitudinal relaxation time (T_1) and a transverse relaxation time (T_2). In rigor muscle at least two-component T_2-relaxation behaviour is observed. T_{2s} and T_{2l} stand for the T_2 with the shortest and longest times respectively and P_{2s} for the population relative to T_{2s}. Highly significant relationships were found between P_{2s}, T_{2l} and some other characteristics such as pH measured 30 min *post mortem* (pH_{30}), reflectance and cooking yield while T_{2s} was correlated only with pH_{30}.[6] Many NMR studies have since been conducted to assess meat quality.[7, 8-10]

What do the water compartments correspond to? Working on pig muscles, Fjelkner-Modig & Tornberg[11] and Tornberg *et al.*[12] identified three water compartments; extracellular water, water in myofibrils and reticulum, and water in interaction with macromolecules. The longest T_2 was regarded as corresponding to "free" or "expelled" water. The main (80%) fraction of water was considered as being held mainly by the myofibrils. This histological picture of a compartmentation of water between intra- and extra-cellular domains is attractive yet there is little evidence to support this concept; structural microheterogeneity being sufficient to explain the non-exponential relaxation[13]. Also, the population associated with the different relaxation times depends on the exchange rates between the different water compartments.[14,15] MRI provides morphological images that can be associated with parametric images of relaxation times, or diffusion in the tissue on a supramolecular scale. Diffusion MRI is a well established tool[16] for non-invasive investigation of structure in muscle. The success of diffusion magnetic resonance imaging is deeply rooted in the powerful concept that, during their random, diffusion-driven displacements, molecules probe tissue structure at a microscopic scale well below the usual image resolution. As diffusion is truly a three-dimensional process, molecular mobility in tissues may be anisotropic. With diffusion tensor imaging (DTI), diffusion anisotropic effects can be fully extracted, characterised, and exploited, providing even finer details of tissue microstructure.

The fibre tract direction is estimated in each voxel after diagonalisation of tensor matrix $\mathbf{D}$. The unit eigenvector related to the largest eigenvalue gives the direction of the tract. The determination of principal fibre directions in muscle tissue contributes substantially to an understanding of water mobility. The apparent diffusion coefficient axially ($D_{//}$) and radially ($D_{\perp}$) can be measured. The $D_{//}$ values measured in beef muscle 24 h *post mortem* were always greater than $D_{\perp}$. The ratio $\dfrac{D_{//}}{D_{\perp}}$, which is an anisotropy index of the water movement, was 1.9. With DTI it is possible to probe the influence of intracellular diffusional barriers[17] during the *post mortem* structural changes.[18]

On the T_2-weighted images obtained by suppression of fat[9], hypersignals were detected in specific locations (Figure 2). The fibre axis is perpendicular to the image plane.

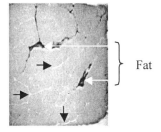

Figure 2   *Fat suppressed image from semitendinosus muscle of beef. The fat appears in black. The hyperintense network characterises free water (black arrow).[9]*

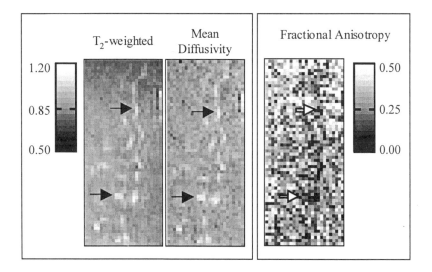

Figure 3   *DTI highlighting accumulation of free water in bovine semitendinosus muscle: T_2 weighted image resulting from the multivariate linear regression of the diffusion weighted images, mean diffusivity scale units in $10^{-3}mm^2s^{-1}$ and fractional anisotropy.[19]*

DTI have confirmed the free water accumulation which diffuses more freely and isotropically than in the rest of the muscle (Figure 3). The fractional anisotropy is another anisotropy index estimated from the diffusion tensor. A null value corresponds to an isotropic diffusion. These results agree with the work of Offer and Cousins[20] who showed by optical microscopy that an interstitial space appears *post mortem* between fascicles of muscle fibres. These voxels correspond to free water exuded into extracellular gaps. These results underline the usefulness of diffusion tensor measurements to characterise muscle structure and help understand the mechanisms of *post mortem* water exudation.[19]

2.2 Tenderness

Tenderness is a major quality factor in meat and is therefore limiting for consumer acceptance.[21] Toughness of meat depends on the connective tissue, the state of the myofibrillar structure and the structural interactions between fibres and the extracellular matrix. The contribution of the amount (or concentration) of intramuscular connective tissue to meat toughness has long been recognised, the connective tissue content being determined by chemical methods, as reviewed by Purslow.[22] In contrast, the role of the spatial distribution of connective tissue in meat quality is still unknown. Muscles with similar connective tissue contents but with different arrangements of connective tissue (orientation, thickness, length) may exhibit different textural properties, such as tenderness, because of the resulting differences in the resistance of the connective network to deformation during mastication and in the mechanical response of the connective tissue to the temperature increase during cooking. Using conventional MRI techniques, direct detection of the connective tissue proton signal is challenging because of the short T_2 relaxation time related to macromolecular protons.[23] During the echo time, spins of the connective tissue protons fully relax and the corresponding NMR signal becomes undetectable. Secondly, efficient excitation of short-T_2 species requires radio-frequency pulses of short duration. Such pulses exhibit poor selectivity in the frequency domain, which limits minimum slice thickness. In heterogeneous samples, inherent magnetic field variations occur owing to the coexistence of two adjacent phases with different magnetic susceptibilities. Because of its low water content, the magnetic susceptibility of collagen-rich connective tissue differs from that of soft tissue.[24] Quantitative assessment of susceptibility effects was performed by mapping T_2^* from the time course of magnitude using multiple Gradient Echo sequence.[25, 26] Comparison with histological pictures indicates that these T_2^* maps exhibit the overall organisation of the primary perimysium at the scale of the whole muscle (Figure 4). The distinct perimysial organisation shown between the Gluteo biceps and *Pectoralis profundis* muscles illustrates the potential of magnetic resonance imaging for characterising muscle connective tissue structure.[27]

MRI / 4.7T   **Histology**

Spatial resolution of
140 x 140 μm²

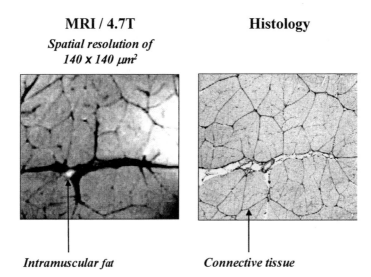

Intramuscular fat   **Connective tissue**

Figure 4   *Bovine* biceps femoris *image underlining the high correspondence between high resolution image and histology (10 μm thick frozen section stained using red Sirius.*

2.3 Fat

For health reasons, fat content is a major consumer concern. To address this concern a correct description of food products is needed, specifically the amounts of their different ingredients. NMR relaxometry[28,29] or spectroscopy[30,31] allows the determination of fat content with great accuracy. The NMR results are always closely correlated with the reference chemical methods. Mitchell *et al.*[32] used the chemical shift difference between water and fat signals to determine the fat content in small birds (in the weight range 93 to 622 g) after slaughter. The correlation between NMR and chemical composition was greater than 0.9. Difficulties in using NMR arise from the fact that measurement depends on the homogeneity of the magnetic field, and the use of this technique is not well suited to large sample sizes. Magnetic resonance imaging (MRI) is a potential alternative tool for examining fat distribution non-invasively and quantitatively, because of the intrinsic contrast due to the different NMR properties of water and lipids.[9] The first MRI study on live animals was carried out by Foster's research group in Aberdeen[33,34] for the measurement of adipose tissue in pigs. More recently, Köver *et al.*[35] performed measurements *in vivo* on commercial broiler chickens. The birds were scanned at six different ages ranging from 6 to 20 weeks to follow the time course of volume of abdominal fat, total body fat and pectoral muscles. The extracellular intramuscular fat 3D distribution in meat was also determined in different cattle breeds (Figure 5). MRI can be very useful in genetic selection and stock breeding.

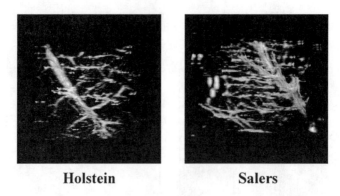

Holstein   **Salers**

Figure 5   *3D images of fat network in muscle from Holstein and Salers cows.*

3 AUTHENTICATION

Products with origin identification are usually high-priced and generate higher profits for producers than ordinary products. For consumers faced with an extensive choice of food commodities, authenticity is a guarantee of safety and eating quality. However, product quality control requires enforcement.

NMR methods can offer powerful tools to enforce food regulations as they can be used to distinguish between different production area, different feeding diets and different technological processes including freezing.

3.1 Geographical origin and diet

Geographical production area, feeding diet and animal breed are the principal characteristics that have to be taken into account. The animal breed can usually be identified using DNA markers. High-resolution gas-phase chromatography coupled with mass spectrometry (HRGC-MS), Isotope ratio mass spectrometry (IRMS) and high-resolution ^{2}H and ^{13}C Nuclear Magnetic Resonance (NMR) spectroscopy are accurate and robust techniques. Stable isotope ratio measured on the whole sample by mass spectrometry and the site-specific ratio determined by NMR are the main isotopic techniques used to characterise the geographical origin of food products. They have been successfully used for the geographical determination of oils, fruit juices and wines.[36] ^{13}C NMR spectra exhibit specific resonance signals characteristic of the unsaturated carbons of polyunsaturated or mono-unsaturated fatty acids.[37] The variation of signal intensities characterizes the feeding diet.[38-40]

The extracted volatile components and the fatty acid composition shared a relationship with diet, while water ^{18}O and ^{2}H enrichment were more closely related to geographical origin. The classification of the samples is achieved using pattern recognition and multivariate calibration and prediction methods. The link between the breeding site and diet markers for meat products can be characterised using these different methods separately or jointly. From ^{2}H and ^{13}C NMR data, discrimination between intensive and extensive pig production was achieved with 95% of well classified animals.

Meat was taken from Charolais bull calves bred in two distinct INRA (Institut National de la Recherche Agronomique) sites differing in their geographical location and altitude: Le Pin, located in Normandy and referred to as *Plain* (altitude 50 m), and Theix, located in the Massif Central referred to as *Mountain* (altitude 950 m). Animals were fed with either grass or maize silage. The production site and feeding diet were determined from NMR and ^{18}O IRMS data (Table 1).

Table 1 *Classification of animals according to diet × geographical origin.*

Origin		Allocation				Classification
Diet		Grass		Maize silage		
	Site	Plain	Mountain	Plain	Mountain	
Grass	Plain	6	-	-	-	100 %
	Mountain	-	6	-	-	100 %
Maize silage	Plain	-	-	5	1	83 %
	Mountain	-	-	1	5	83 %
Total						92 %

3.2 Freezing-thawing

Freezing is currently used for extending the shelf life of meat by inhibiting microbiological growth. However, the price of fresh or chilled meat is higher than that of frozen-thawed meat. Although less sensitive than many other spectroscopic techniques, MRI has already been used to measure the effects of freezing in beef, lamb and pork meat.[41] The effect of different freezing methods on trout muscle has been investigated.[42,43] The diffusion coefficient perpendicular to the muscle fibre and the T_2 were found to be the NMR parameters most sensitive to the freezing storage period (Figure 6). The variations in these parameters agree with histological observations such as fibre separation observed from the first day of frozen storage and the breakage of connective tissue and torsion of the fibres shown at day 41.[43]

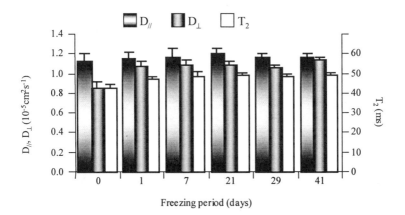

Figure 6 *NMR parameters for water in unfrozen and frozen-thawed trout for different freezing periods.*

3.3 Brine

Salt (sodium chloride), a substance essential for life processes is the second most often used food additive. Salt is added as a flavouring or flavour enhancer, as a preservative and as an ingredient contributing to desirable textural characteristics of meat products.

Excessive dietary sodium intake is believed to contribute to hypertension and the development of cardiovascular disease. Health authorities advocate reducing sodium intake in the general population and especially certain at-risk consumer groups. However, reducing salt levels may seriously impair the inherent bacteriological stability of the products, and be detrimental to their overall quality.

There are a number of methods for determining the total ion concentration in food products. Except for NMR, no method is able to measure *in situ* the ratios $\dfrac{[Na^+]_{bound}}{[Na^+]_{free}}$ and $\dfrac{[Cl^-]_{bound}}{[Cl^-]_{free}}$. The ^{35}Cl and ^{23}Na ions possess an electric quadrupolar moment (3/2 spin). For the free ions the conditions of extreme narrowing hold. In biological tissue the bound ions undergo static or dynamic quadrupolar effects.[44] The NMR parameters are different according to binding state.[45] The quantification can be performed using specific sequences.[46] The quantitative NMR data are closely correlated with the chemical method. In addition, for each ^{35}Cl and ^{23}Na ion, the bound/free ion ratio reveals significant differences according to technological processing method[47] (Figure 7).

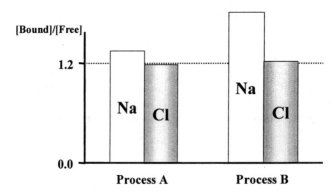

Figure 7 *Diagram of* $\dfrac{[Na^+]_{bound}}{[Na^+]_{free}}$ *and* $\dfrac{[Cl^-]_{bound}}{[Cl^-]_{free}}$ *ratios according to two technological processes.*

3.4 Drying

Food characteristics are greatly influenced by moisture content. Texture, stability and microbiological growth depend mainly on the availability of water. Methods for processing or stabilising solid foods involve coupled water and heat transfer. They must be optimised to reduce operating costs and maximise product quality. Internal water migration is a function of chemical composition and structure, and drives the overall water transfer. The accurate determination of water distribution and water diffusivity coefficient (D) are very important. This coefficient D varies with water content. It can be derived from the time course of the moisture profile measured by NMR.[48] During the drying process the shrinking of a pork meat sample was non-uniform. This spatial heterogeneity required extracting the profiles of water content from a central band of a 2D image. In these conditions the spatial resolution was 50 μm and temporal resolution 15 min.

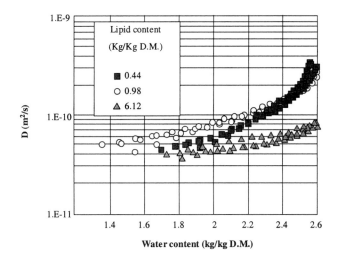

Figure 8 *Diffusivity measured by NMR according to water content for meat samples with different fat content*

The effects of lipid content, fibre direction and drying temperature were studied. From the water content profiles, the fibre orientation has no effect on D, while the water diffusivity strongly depends on the water and lipid contents (Figure 8). The lipid content has a negative effect on D while the temperature induces an increase in D values for low water content and a decrease for high water content.[49][50]

4 CONCLUSION

NMR studies in spectroscopy and imaging can afford a better understanding of water interactions in meat structure underlying meat quality. The NMR analytical methods are highly sensitive, accurate and robust but are often expensive and complicated to implement in the meat industries, limiting their applicability. Magnetic Resonance Imaging can offer a reference method for solving specific problems. The domains where nuclear magnetic resonance could be used are animal selection and authenticity monitoring. For genetic selection, the number of animals could be reduced, owing to the non-invasive character and the great accuracy of NMR. The control of authenticity of meat product corresponds to a consumer demand for a guarantee of safety and eating quality. Inspections are performed on small sample numbers and few methods are available except for mass spectrometry and magnetic resonance spectroscopy. In view of the marked industrialisation of meat processing, NMR may be useful for optimising technological processes such as brining, drying and freezing.

References

1   D.J. Etherington and T.J. Sims, *Journal of the Science of Food and Agriculture,* 1981, **32**, 539.
2   G. Offer and P. Knight, 'The structural basis of water-holding in meat. Part 2: Drip losses' in *Development in Meat Science*, eds., R.A. Lawrie, Elsevier Science Publishers, London, 1988, Vol. 4, pp. 172-243.

3   G. Offer and P. Knight, 'The structural basis of water-holding in meat. Part 1: general principles and water uptake in meat processing' in *Development in Meat Science*, eds., R.A. Lawrie, Elsevier Science Publishers, London, 1988, Vol. 4, pp. 63-171.

4   J.P. Renou, J. Kopp, P. Gatellier, G. Monin, and G. Kozak-Reiss, *Meat Science.,* 1989, **26**, 101.

5   J. Brøndum, L. Munck, P. Henckel, A. Karlsson, E. Tornberg, and S.B. Engelsen, *Meat Science,* 2000, **55**, 177.

6   J.P. Renou, G. Monin, and P. Sellier, *Meat Science.,* 1985, **15**, 225.

7   Borowiak P., J. Adamski, K. Olszewski, and J. Bucko, *32th European Meeting of Meat Research Workers, Ghent,* 1986, p. 467.

8   R.J. Brown, F. Capozzi, C. Cavani, M.A. Cremonini, M. Petracci, and G. Placucci, *Journal of Magnetic Resonance,* 2000, **147**, 89.

9   W. Laurent, J.M. Bonny, and J.P. Renou, *Journal of Magnetic Resonance Imaging,* 2000, **12**, 488.

10   H.C. Bertram, H.J. Andersen, and A.H. Karlsson, *Meat Science,* 2001, **57**, 125.

11   S. Fjelkner-Modig and E. Tornberg, *Meat Science,* 1986, **17**, 213.

12   E. Tornberg, A. Andersson, A. Göransson, and G. Von Seth, 'Water and fat distribution in pork in relation to sensory properties.' in *Pork Quality, Genetic and Metabolic Factors*, eds., E. Puolanne and D. Demeyer, CAB International, Townbridge, 1993, pp. 239-263.

13   A. Traoré, L. Foucat, and Renou J.P, *Biopolymers,* 2000, **53**, 476.

14   B.P. Hills, *Molecular Physics,* 1992, **76**, 509.

15   B.P. Hills, *Molecular Physics,* 1992, **76**, 489.

16   D. Le Bihan, J.F. Mangin, C. Poupon, C.A. Clark, S. Pappata, N. Molko, and H. Chabriat, *Journal of Magnetic Resonance Imaging,* 2001, **13**, 534.

17   S.T. Kinsey, B.R. Locke, B. Penke, and T.S. Moerland, *NMR in Biomedicine,* 1999, **12**, 1.

18   L. Foucat, S. Benderbous, G. Bielicki, M. Zanca, and J.P. Renou, *Magnetic Resonance Imaging,* 1995, **13**, 259.

19   J.M. Bonny and J.P. Renou, *Magnetic Resonance Imaging,* 2002, **20**, 395.

20   G. Offer and T. Cousins, *Journal of the Science of Food and Agriculture,* 1992, **58**, 107.

21   G. Monin, *Meat Science,* 1998, **49**, 231.

22   P. P. Purslow, *45th ICOMST*, 1999, p. 210.

23   H.T. Edzes and E.T. Samulski, *Journal of Magnetic Resonance,* 1978, **31**, 207.

24   J.F. Schenck, *Med Phys,* 1996, **23**, 815.

25   S. Posse and W.P. Aue, *Journal of Magnetic Resonance,* 1990, **88**, 473.

26   D.A. Yablonskiy, *Magn Reson Med,* 1998, **39**, 417.

27   J.M. Bonny, W. Laurent, and J.P. Renou, *Magnetic Resonance Imaging,* 2000, **18**, 1125.

28   J.P. Renou, J. Kopp, and C. Valin, *Journal of Food Technology,* 1985, **20**, 23.

29   C.A. Toussaint, F. Medale, A. Davenel, B. Fauconneau, P. Haffray, and S. Akoka, *Journal of the Science of Food & Agriculture. 2002. 82: 2, 173-178. 22 Ref..*

30   L. Foucat, J.P. Donnat, F. Humbert, G. Martin, and J.P. Renou, *Journal of Magnetic Resonance Analysis,* 1997,108.

31   J.P. Renou, A. Briguet, P. Gatellier, and J. Kopp, *International Journal of Food Science and Technology,* 1987, **22**, 169.

32   A.D. Mitchell, P.C. Wang, R.W. Rosebrough, T.H. Elsasser, and W.F. Schmidt, *Poultry Science,* 1991, **70**, 2494.

33   P.A. Fowler, M.F. Fuller, C.A. Glasbey, G.G. Cameron, and M.A. Foster, *American Journal of Clinical Nutrition,* 1992, **56**, 7.

34   M.F. Fuller, P.A. Fowler, G. McNeill, and M.A. Foster, *Journal of Nutrition,* 1994, **124**, 1546S.

35   G. Köver, R. Romvári, P. Horn, E. Berényi, J.F. Jensen, and P. Sørensen, *Acta Veterinaria Hungarica,* 1998, **46**, 135.

36   G.J. Martin and M.L. Martin, 'Stable isotope analysis of food and beverages by nuclear magnetic resonance.' in *Annual Reports on NMR Spectroscopy.,* eds.1995, Vol. 31, pp. 81-104.

37   S. Ng and W.L. Ng, *Journal of the American Oil Chemists Society,* 1983, **60**, 1266.

38   M. Bonnet, C. Denoyer, and J.P. Renou, *International Journal of Food Science and Technology,* 1990, **25**, 399.

39   S.C. Cunnane, T. Allman, J. Bell, M.J. Barnard, G. Coutts, S.C. Williams, and R.A. Iles, 'In vivo fatty acid analysis in humans and animals using carbon-13 nuclear magnetic resonance spectroscopy.' in *Human Body Composition,* eds.1993, Vol. 60, pp. 355-358.

40   T.W. Fan, A.J. Clifford, and R.M. Higashi, *Journal of Lipid Research,* 1994, **35**, 678.

41   S.D. Evans, K.P. Nott, A.A. Kshirsagar, and L.D. Hall, *International Journal of Food Science and Technology,* 1998, **33**, 317.

42   K.P. Nott, S.D. Evans, and L.D. Hall, *Magnetic Resonance Imaging,* 1999, **17**, 445.

43   L. Foucat, R.G. Taylor, R. Labas, and J.P. Renou, *American Laboratory,* 2001, **33**, 38.

44   J.P. Renou, S. Benderbous, G. Bielicki, L. Foucat, and J.P. Donnat, *Magnetic Resonance Imaging,* 1994, **12**, 131.

45   H.J. Berendsen and H.T. Edzes, *Annals of the New York Academy of Sciences,* 1973, **204**, 459.

46   E.M. Shapiro, A. Borthakur, R. Dandora, A. Kriss, J.S. Leigh, and R. Reddy, *Journal of Magnetic Resonance,* 2000, **142**, 24.

47   L. Foucat, Donnat J.P., and Renou J.P. ^{23}Na and [35]Cl NMR studies of the interactions of sodium and chloride ions with meat products.' in *Magnetic Resonance in Food Scienc,* eds., G.A. Webb, P.S. Belton, A.M. Gil, and D.N. Rutledge, Royal Society of Chemistry, in press

48   L. Foucat, M. A. Ruiz-Cabrera, J. D. Daudin, C. Bonazzi, and Renou J.P., *11[e] Rencontres Scientifiques Et Technologiques Des Industries Alimentaires AGORAL, Nantes,* 1999, p. 391.

49   M. A. Ruiz-Cabrera, P. Gou, L. Foucat J. P. Renou, and J. D. Daudin, *13th International Drying Symposium 2002, Beijing, China.*

50   M.A. Ruiz-Cabrera, P. Gou, L. Foucat, J.P. Renou, and J.D. Daudin, *Meat Science,* submitted.

¹H NMR RELAXATION STUDY OF AMYLOPECTIN RETROGRADATION

I.A Farhat, M.A Ottenhof, V. Marie and E. de Bezenac

Division of Food Sciences, School of Biosciences, University of Nottingham, Loughborough, LE12 5RD, UK

1 INTRODUCTION

The changes occurring in starch and starch based materials during heat and moisture treatment, cooling and subsequent storage are of major scientific and industrial interests. From a nutritional point of view, the digestibility of native starches is greatly increased during cooking through the application of heat and/or shear in the presence of water as the starch polysaccharides become more susceptible to enzyme hydrolysis. During post-cooking cooling and the early part of storage (minutes timescales), amylose gelation, resulting from the ordering and association of this mainly linear polysaccharide, occurs. This process, also referred to as amylose retrogradation, leads to significant rheological changes. Retrograded amylose exhibits low amylase digestibility.

During the storage of starch based systems in the rubbery state (e.g. at water contents in excess of 20% wet basis at positive temperatures), for example in baked goods, cooked pastas and rice, etc., amylopectin retrogradation occurs. This is a recrystallisation process, the kinetics of which follows similar patterns to those established for many synthetic polymers.[1] Amylopectin retrogradation is believed to play a major role in the staling of starchy foods leading to changes to the perception (harder texture and dryer mouth feel) and the nutritional value (decreased digestibility).[2] The decreased amylasis of retrograded starches has been exploited industrially to produce a range of resistant starch products, a functional ingredient with nutritional and health benefits, comparable to fibres.

Monitoring starch retrogradation, particularly in intermediate moisture foods (between 20% and 60% wet basis), is thus an important aspect in determining the shelf-life of starch based systems and the efficiency of resistant starch production.

A wide range of techniques have been developed and employed to study retrogradation. Wide angle x-ray diffraction is one of the most established methods.[3,4,5] It was utilised as early as 1912 by Katz who suggested that bread staling is linked to starch re-crystallisation in his review published in 1928.[6] Other widely used techniques include differential scanning calorimetry,[7,8,9] rheology[10,11] and texture[12] measurements and infrared spectroscopy.[13,14,15] The use of NMR to study starch retrogradation is relatively recent. A range of NMR techniques have been adopted, these include time domain ¹H relaxometry,[1,16,17,18] ¹H cross-relaxation[19,20,21,22] and high resolution ¹³C solid-state[23,24,25] techniques to monitor the changes in the mobility of water and the starch biopolymers during retrogradation.

In this chapter, the use of proton NMR relaxometry to monitor amylopectin retrogradation is described and the advantages of this approach when compared to other more widely used techniques is discussed with particular emphasis on practical issues.

2 MATERIALS AND METHODS

Waxy maize starch (wms) containing a minimum of 98% amylopectin (National Starches and Chemicals Co., UK) was extruded into non-expanded ribbons through the 1mm x 30mm slit die of a Clextral BC-21 co-rotating twin-screw extruder operating at 120°C with a die temperature of 90°C. The post-extrusion water contents, determined by overnight drying at 105°C, were 32±1% wet weight basis (wwb). The samples were sealed in 10mm NMR tubes, stored and analysed at 22±1°C.

NMR experiments were performed using a MARAN (Resonance Instruments) spectrometer operating at 23 MHz. FID data were acquired using the solid-echo pulse sequence with an inter 90° pulse spacing of 10μs and a dwell time of 2μs. CPMG decays acquired at different 90°-180° pulse spacing (τ) were fitted to a continuous distribution of exponentials using Resonance WinDXP software with the shortest T_2 value set to 2τ.

3 MONITORING AMYLOPECTIN RETROGRADATION

The occurrence of amylopectin retrogradation i.e. recrystallisation on storage was assessed using one of the most established physical techniques in the retrogradation literature, namely wide angle x-ray diffraction. As expected, the diffractogram acquired shortly after extrusion showed a diffuse pattern typical of amorphous starch confirming the efficiency of the extrusion process in disrupting thermo-mechanically the partially crystalline starch granules. Crystallisation of amylopectin during storage was evident from the appearance of the sharp diffraction peaks associated with crystalline A-type starch.

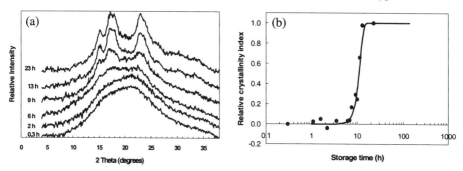

Figure 1   *WAXS diffractograms showing the crystallisation of amylopectin during the ageing of wms in the rubbery state (a) and the dependency of crystallinity calculated from the diffractograms on ageing at 23°C (b)*

The increase of crystallinity with time was quantified from the increased spectral resolution as described elsewhere .[26] The results are displayed in Figure 1b. Although the underlining increase in crystallinity index on storage is clear, the results exhibited a high degree of scatter. Such scatter is typical of the method and is partly due to the quantification of crystallinity from the raw x-ray diffractograms.

The proton relaxation behaviour of the same sample was also monitored as a function of ageing time. The FIDs exhibited a relatively small but systematic increase of the contribution of the rigid, solid-like component, which also decayed more rapidly (Figure 2a). These observations reflect the decreased mobility of the amylopectin chains as they become involved in crystallites. It is expected that the decay rate of the slowly decaying component of the FID is greatly affected by magnetic field inhomogeneity and thus does not strictly report on the mobility of the mobile fraction of the material under observation. Therefore, spin-echo decays were measured using the CPMG pulse sequence.

The CPMG decays showed a much more pronounced dependency on ageing with increased decay rates at longer storage times (Figure 2b).

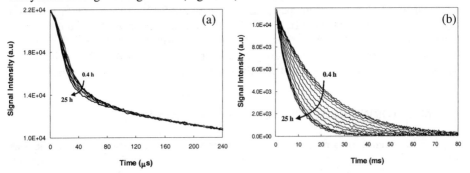

Figure 2   *Effect of amylopectin retrogradation on the FID (a) and the CPMG (b) decays.*

The relaxation spectra were obtained by fitting the spin-echo decays to a continuous distribution of exponentials. The decays acquired at a pulse spacing $\tau=200\mu s$, showed 1 main relaxation time distribution (Figure 3a). This is described in section 3.

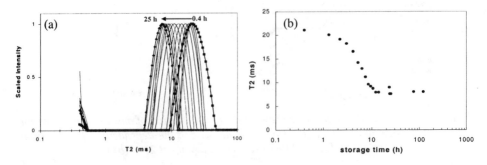

Figure 3   *Effect of amylopectin retrogradation on the CPMG relaxation spectrum (The maximum intensity values were scaled to 1) (a) and the main relaxation time (b)*

As the retrogradation progressed, the characteristics of the T_2 distribution changed. The width decreased from ~39 ms after 0.4 h of storage down to ~10 ms after 25 h while the main T_2 of the distribution decreased from ~21 ms down to ~7 ms over the same storage period (Figure 3). Since the mobility of amylopectin is expected to dramatically decrease as a result of the crystallisation of its amylopectin A-chains, the decrease in T_2 would therefore be expected since the spin-spin relaxation rate of water does, to a large extent, mirror that of the matrix due to proton exchange and cross-relaxation processes. However, the narrowing of the relaxation spectrum is somehow counter-intuitive since the molecular heterogeneity increases as the system evolves from a relatively homogeneous amorphous state to a heterogeneous partially crystalline state through retrogradation. An attempt to interpret this observation is made in section 4.

The application of the well-known Avrami kinetic to the spin-spin relaxation rates monitoring the progress of retrogradation was described in detail elsewhere[27] and is summarised in Figure 4.

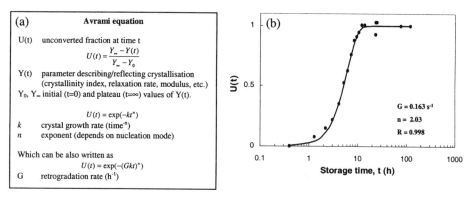

Figure 4   *Description of the Avrami kinetics equation (a) and its application to the CPMG T_2 to quantify the kinetics of amylopectin retrogradation (b)*

4 PROBING WATER MOBILITY

It is well known that staling involves, in addition to starch retrogradation, several other physical processes. Water is involved in most of them. These include, in the case of baked systems, water redistribution among the various constituents (crystalline and amorphous fractions of amylopectin, amylose, gluten, pentosans, etc.), water redistribution between the crumb and the crust and change of the overall moisture content.
In a previous study on starch-gelatin mixtures,[28] we have demonstrated that the crystallisation of amylopectin, particularly to the relatively anhydrous A-type polymorph, leads to an increase in the amount of water associated with both the amorphous fraction of amylopectin and the protein.

The effect of amylopectin retrogradation on the dynamics of water over a range of timescales was studied by examining the spin-spin relaxation spectra obtained from CPMG decays recorded at increasing τ spacings ranging from 0.05 to 2 ms on fresh and retrograded samples (Figure 5).
In addition to the main T_2 distribution described above, the relaxation spectra showed the beginning of a low T_2 distribution centred around 0.2ms. However since this component

of the spectra was not fully recorded, it is difficult to assess the effect of retrogradation has on it and therefore no attempt to assign/interpret this component was made. The main distribution exhibited the behaviour described above (Figure 3) in terms of a shift to lower T_2 and narrowing of the distribution width as the sample retrograded. Another noticeable pattern is the narrowing of the T_2 distribution as the observation time increased. In order to attempt to understand these observations, the average displacement ($<r>$) of a water molecule in the time scale of interest ($t=2\tau$, echo time) was estimated using the relationship $<r> = (2Dt)^{1/2}$. A diffusion coefficient value of $D=0.3 \times 10^{-9}$ $m^2.s^{-1}$ was used.[29] $<r>$ ranged between ≈ 0.2 μm and ≈ 1.2 μm for τ values between 0.05 ms and 2 ms. These distance scales are much larger than the expected thickness of crystalline lamellae (typically 10 nm) justifying the observation of one main T_2 distribution for water over all τ and ageing time values as a result of the diffusive exchange between the various water populations.[30,31,32] The extremely wide relaxation time distribution at $\tau=0.05$ ms, particularly in the fresh sample, implies a heterogeneity of the water environments over a sub-micrometer distance scale. The origin of such heterogeneity is likely to be remnants of partially swollen starch granules. The narrowing of the T_2 distributions on ageing suggests the occurrence of a structure, uniform over a micrometer distance scale made of sub-micrometer crystallites dispersed in an amorphous rubbery matrix. This suggestion is compatible with the uniformity of these samples when examined by light microscopy.

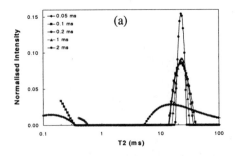 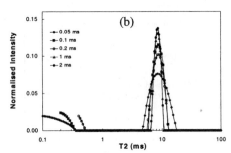

Figure 5   *CPMG spin-spin relaxation spectra acquired at different pulse spacings (τ) after 0.4 h (a) and 24 h (b) of storage. The intensities were normalised to the area of the main relaxation distribution*

5 ADVANTAGES OF ^{1}H NMR IN RETROGRADATION STUDIES

From our experience in studying starch retrogradation in model and real food systems, ^{1}H relaxometry using common low field spectrometers such as the Minispec (Bruker) or the Maran (Resonance) and others, offers several advantages.

In addition to providing an insight into the change in molecular dynamics occurring during starch retrogradation in particular and staling in general (including water re-distribution, moisture migration, etc.), the approach has many important practical advantages over other techniques more established in the field. These are summarised in Table 1 and discussed below.

Table 1 *Comparison of the main physical techniques used in starch retrogradation studies in intermediate and low water content systems based on our experience. The techniques are listed in a descending order of popularity in the retrogradation/staling literature*

Technique	Advantages	Disadvantages
Wide Angle X-Ray Diffraction	•Established •Direct measurement of crystallinity •Direct information on polymorphism •"non-destructive" [1]	•Surface technique (mostly) •Quantification of crystallinity [2] •Difficult control of moisture/temperature •Slow (up to hours) depending on signal/noise
Differential Scanning Calorimetry	•Established •Monitors molecular/crystalline order •Excellent control of moisture content and temperature	•Small sample (typically mg) •Destructive [3] •Slow (minutes-hours [4]) •Reproducibility due to sampling •Cost (equipment and special pans)
Texture analysis and other rheological techniques	•Established •Easy to relate to as a food technologist or a consumer •Cost/availability in the Food Industry	•Poor reproducibility •Difficult control of moisture loss and temperature •Often destructive
Low field time domain ^{1}H NMR	•Control of moisture and temperature •Rapid (seconds-minutes) •Non-destructive •Sample size (several grams) •Monitors the bulk of the sample •Cost/availability [5]	•Interpretation
Solid state high-resolution NMR (mostly ^{13}C CP-MAS)	•Direct measurement of molecular order •Changes in carbohydrate/protein can be studied simultaneously (e.g in baked goods). •Non-destructive •Control of moisture and temperature •Monitors the bulk of the sample	•Quantification is difficult due to overlapping peaks. •Slow (hours [6]) •Cost
mid-infrared (FTIR)	•Monitors molecular order •Simultaneous study changes in carbohydrate/protein (baked goods). •Rapid (seconds-minutes) •Control of moisture loss and temperature •"Non-destructive" [7]	•Small sample size [8] •Quantification [9]

1 Samples may have to be milled for better results.
2 Crystallinity calculations rely on subjective baseline definition. Also polymorphic transformations and occurrence of mixtures of polymorphs depending on retrogradation conditions (temperature, water content, etc.) make quantification protocols relying on crystalline standards impracticable.
3 Unless used in isothermal mode (not readily feasible on DSC). Often sample needs to be suspended in excess water due to the dependence of starch melting temperature and enthalpy on water content.
4 In isothermal mode.
5 Often available in the Food Industry for other applications such as oil/moisture content, solid fat content, droplet size analysis, etc.
6 For ^{13}C experiments in natural abundance.
7 Not strictly, due to small sampling volumes (around 0.1 mm^3).
8 Transmission on thin films (<100 μm), Reflectance on surface (~2 μm depending on wavelength).
9 Quantification of molecular order relies on deconvolution of overlapping absorbencies and numerical line-narrowing procedures. The quantification is water content dependent.

The aspects of sample volume and control of moisture content and temperature are rarely considered. However they are of major importance[1] and thus they are discussed in details in this section.

Samples of approximately 2g were used in this study but low-field instruments capable of measuring spin-echo decays (long dead times) on samples of the order of 50 or even 100g are now readily available. Larger samples can be analysed using equipment designed for MRI. Also of particular importance is the fact that NMR reports on the whole of the sample while several other techniques monitor only the first few micrometers of the surface. These sampling issues are very important since most foods are heterogeneous.

The technique enables a reliable control of temperature (typically ±0.1°C) and moisture controls (NMR tubes can be made airtight using special sealants or through flame sealing). Potential moisture content changes during storage can be continuously monitored by plotting the maximum intensity of the FID signal (at t=0μs) as a function of storage time.

While many techniques can be described as being "non-destructive", they are not strictly so, since they rely on sub-sampling a small aliquot, which can be a few milligrams in the case of Differential Scanning Calorimetry, a thin film when using transmission mid-infrared spectroscopy or less than a gram in the case of solid state NMR. As described above samples for low field NMR studies can be much larger approaching true non-destructiveness.

6 CONCLUSIONS

This paper briefly described an example of studying amylopectin retrogradation in the model system of extruded waxy maize starch. This approach has been extended to study several other model and real starch based foods (baked goods, cooked grains, etc.). Low field time domain ^{1}H NMR offers several advantages over other more established and widely used physical techniques (sampling, non-destructive, speed, moisture/temperature control, etc.). The ability of the technique to monitor molecular dynamics, particularly that of water, over a range of time/distance scales could offer new insights into the effect of retrogradation on mouth-feel and the role of some anti-staling strategies.

References

1. I.A. Farhat, J.M.V. Blanshard and J.R. Mitchell, *Biopolymer*, 2000, **53**, 411-422.
2. I.A. Farhat, J. Protzmann, A. Becker, B. Vallès-Pàmies, R. Neale and S.E. Hill, *Starch-Stärke*, 2001, **53**, 431-436.
3. S.G. Ring, P. Colonna, K.J. L'Anson, M.T. Kalichevsky, M.J. Miles, V.J. Morris and P.D. Orford, *Carbohydrate Research*, 1987, **162**, 277-293.
4. R.D.L. Marsh and J.M.V. Blanshard, *Carbohydrate Polymers*, 1988, **9**, 301-317.
5. P. Cairns, K.J. I'Anson, and V.J. Morris, *Food Hydrocolloids,* 1991, **5**, 151-153.
6. J.R. Katz in: *A Comprehensive Survey of Starch Chemistry*, Ed. R.P. Walton, The Chemical Catalog Company Inc: New York, 1928, **1**, 100-117.
7. C.G. Biliaderis and D.J. Prokopowich, *Carbohydrate Polymers*, 1994, **23**,193-702.
8. D.K. Fisher and D.B. Thompson, *Cereal Chemistry*, 1997, **74** (3), 344-351.
9. Q. Liu and D.B. Thompson, *Carbohydrate Research*, 1998, **314** (3-4), 221-235.
10. L.H. Aee, K.N. Hie and K. Nishinari, *Thermochimica Acta*, 1998, **322** (1), 39-46.

11. F.B. Ahmad and P.A. Williams, *Biopolymers*, 1999, **50** (4), 401-412.
12. I. Lima and R.P. Singh, *Journal of Food Quality*, 1993, **16** (5), 321-337.
13. B.J. Goodfellow and R.H. Wilson, *Biopolymers*, 1990, **30**, 1183-1189.
14. R.H. Wilson, B.J. Goodfellow, P.S. Belton, B.G. Osborne, G. Oliver and P.L. Russel, *Journal of the Science of Food and Agriculture*, 1991, **54**, 471-483.
15. J.J.G. Van Soest, D. Dewit, H. Tournois, J.F.G. Vliegenthart, *Polymer*, 1994, **35**, 4722-4727.
16. F. Schierbaum, S. Radosta, W. Vorwerg, V.P. Yuriev, E.E. Braudo and M.L. German, *Carbohydrate Polymers*, 1992, **18**, 155-163.
17. C.H. Teo and P. Seow, *Starch-Starke*, 1992, **44** (8), 288-292.
18. D. Le Botlan and P. Desbois, *Cereal Chemistry*, 1995, **72** (2), 191-193.
19. Y.W. Junshi and M.E. Thomas, *Carbohydrate Polymers*, 1993, **20**, 51-60.
20. J.Y. Wu, R.G. Bryant and T.M. Eads, *Journal of Agricultural and Food Chemistry*, 1992, **40**, 449-455.
21. J.Y. Wu and T.M. Eads, *Carbohydrate Polymers*, 1993, **20**, 51-60.
22. Y. Vodovotz, E. Vittadini and J.R. Sachleben, *Carbohydrate Research*, 2002, **337**, 147-153.
23. K.R. Morgan, R.H. Furneaux and R.A. Stanley, *Carbohydrate Research*, 1992, **235**, 15-22.
24. K.R. Morgan, J. Gerrard, D. Every, M. Ross and M. Gilpin, *Starch-Starke*, 1997, **49**, 54-59.
25. A.L.M. Smits, F.C. Ruhnau, J.F.G. Vliegenthart and J.J.G van Soest, *Starch-Starke*, 1998, **50**, 478-483.
26. I.A. Farhat, 1996, Phd Thesis, University of Nottingham.
27. I.A. Farhat and J.M.V. Blanshard in: *Bread Staling*, Eds. Y. Vodovotz and P. Chinachoti, CRC Press, 2000, 163-172.
28. I.A. Farhat, Z. Mousia and J.R. Mitchell, *Polymer*, 2001, **42**, 4763-4766.
29. I.A. Farhat, E. Loisel, P. Saez Rojo, W. Derbyshire and J.M.V. Blanshard, *International Journal of Food Science and Technology*, 1997, **32**, 377-387.
30. P.J. Lillford, A.H. Clark and D.V. Jones, *A.C.S. Symp. Series*, 1980, **127**, 177.
31. P.S. Belton and B.P. Hills, *Molecular Physics*, 1987, **61** (4), 999-1018.
32. B.P. Hills, S.F. Takacs and P.S. Belton, *Food Chemistry*, 1990, **37**, 95-111.

^{23}Na AND ^{35}Cl NMR STUDIES OF THE INTERACTIONS OF SODIUM AND CHLORIDE IONS WITH MEAT PRODUCTS

L. Foucat, J.P. Donnat and J.P. Renou

Structures Tissulaires et Interactions Moléculaires, SRV, INRA-Theix, 63122 Saint-Genès Champanelle, France

1 INTRODUCTION

Sodium chloride is widely used in the meat industry to improve meat quality characteristics such as water holding capacity, in addition to its primary function as an inhibitor of microbial spoilage flora growth. However, it is claimed that excessive salt consumption is linked to cardiovascular disease. Reduction of salt content in brined products is therefore recommended in public health policies, but this reduction must not impair either organoleptic qualities or preservation properties. To achieve a safe salt reduction, a better understanding of the role of Na^+ and Cl^- ions is necessary. In particular, it is essential to quantify and characterise for each ion the "bound" and "free" fractions in relation to the microbiological quality and the salt taste of the product.

^{23}Na and ^{35}Cl NMR spectroscopy offers a unique opportunity to access non-invasively the distribution and state of Na^+ and Cl^- ions in tissues.[1,2,3] To characterise the dynamics of these two quadrupolar nuclei (I=3/2), single quantum (SQ) and multiple quantum filter (MQF) acquisitions were used. MQF NMR spectroscopy is particularly well suited to the study of ordered biological tissues. An MQ filter deletes the NMR signal belonging to isotropically reorienting ions, leaving only the signal from ions exhibiting biexponential decay (bound or entrapped ions).

Preliminary results of the application on two different salted meat products (pork meat and smoked salmon) are reported to illustrate the NMR and data treatment procedures used.

2 MATERIALS AND METHODS

2.1 Samples

2.1.1 Pork meat. Fresh pork loin samples (*longissimus dorsi* muscle) were processed and supplied by a local abattoir (ADIV, Clermont-Ferrand). A first loin sample (A) was brined with fine salt (30 g/kg of meat) and saltpetre (0.3 g/kg of meat), vacuum-packed in a plastic bag and preserved at 4°C. A second loin sample (B) was non-salted and preserved under vacuum at 0°C. After 15 days of storage in a cold room, loins A and B were minced (6 mm diameter mincer), and mixed in different proportions to obtain samples with salt contents ranging from 0 to 30 g/kg of meat. Two other samples were prepared separately, in the same way as described above, with 50 and 70 g of salt per kg of meat. For each salt content, NMR experiments were repeated on two different samples.

2.1.2 Smoked salmon. Smoked salmon fillets were obtained from two French industrial manufacturers (A and B) that use different salting process (dry salt for A; combined dry salt and brine injection for B). A cylinder of smoked salmon (10 mm in diameter) was cut from the fillet and inserted into the NMR tube. To investigate the heterogeneity of the salt distribution in smoked salmon, two samples were extracted from the thick and thin parts of the fillet. For each salting process and each sample location, NMR experiments were repeated on a series of five fillets.

2.2   NMR experiments

All the NMR experiments were performed at 106 MHz (^{23}Na) and 39 MHz (^{35}Cl) on a Bruker DRX 400 (9.4 T). A 10 mm ^{23}Na/^{35}Cl double tune NMR probe was devised for this investigation. Samples were maintained in the spectrometer at 4°C. For the overall NMR acquisitions, recycle times of 200 and 100 ms were used for ^{23}Na and ^{35}Cl nuclei, respectively.

2.2.1 External references. Aqueous solutions of Na$_7$Dy(PPP)$_2$ and DyCl$_3$ were prepared and used as external references (5 mm NMR tube placed inside the 10 mm tube containing the sample studied) for sodium and chloride ion quantification, respectively. The Dy(PPP)$_2$$^{7-}$ and Dy^{3-} ions induced an upfield shift of Na$^+$ (ΔNa = 30 ppm) and a downfield shift of Cl$^-$ (ΔCl = -55 ppm).[4] The quantification was validated on a series of ten sodium chloride solutions, with NaCl concentrations ranging from 1.5 to 59 g/l. The calibration curves are presented in Figure 1. An excellent correlation between true NaCl concentration and those determined by NMR was obtained for both Na$^+$ and Cl$^-$ ions.

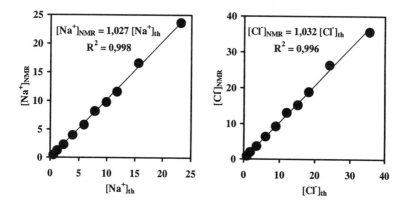

Figure 1   *NMR calibration curves of Na$^+$ and Cl$^-$ ion content (g/l)*

2.2.2 Double quantum filtered (DQF) experiments. The basic multiple quantum-filtered (MQF) pulse sequence is given by:[5]

$$90° - \tau/2 - 180° - \tau/2 - \theta° - \delta - \theta° - \text{acquire} \tag{1}$$

where τ and δ are the multiple quantum creation and evolution times, respectively. The formation of double quantum (DQ) coherences, with $\theta = 90°$, was detected by proper phase cycling.[6] As shown in Figure 2, the DQ filtration places the two relaxing magnetisations in antiphase. The absorption spectrum is the difference of two Lorentzians, of equal areas, but different widths. All deconvolution and peak area analyses were carried out with PeakFit ™ software.

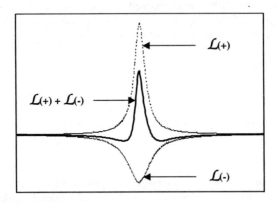

Figure 2 *Example of double quantum filtered* ^{23}Na *spectrum (solid line) from minced pork sample, and the result of the deconvolution in two Lorentzian peaks, $\mathcal{L}(+)$ and $\mathcal{L}(-)$, in antiphase (dashed lines)*

To optimise the response of DQ-filtered experiments, the two time constants of the biexponential sodium decay were determined. For this purpose, a number of DQ-filtered spectra were recorded as a function of the creation time τ (T_2-DQF experiment). The corresponding relaxation time constants, usually denoted fast and slow (T_{2f} and T_{2s}), were then extracted by plotting the peak heights (I) of the processed spectra versus τ (Figure 3) and fitting the points to the function

$$I(\tau) = k \times \left[\exp\left(-\tau/T_{2s}\right) - \exp\left(-\tau/T_{2f}\right) \right] \tag{2}$$

Once the transverse relaxation times T_{2f} and T_{2s} were obtained, the value of the creation time, τ_{max}, which maximizes the DQF signal intensity (Figure 3), was obtained from

$$\tau_{max} = \frac{\text{Ln}\left(T_{2s}^{-1}/T_{2f}^{-1}\right)}{T_{2f}^{-1} - T_{2s}^{-1}} \tag{3}$$

The DQF signals were then quantified relative to the SQ signals, as previously described in the literature.[7] The SQ signal acquisition consists in a spin echo sequence with identical transverse magnetisation evolution time, and similar scan number and receiver gain to those used for the DQF acquisition.

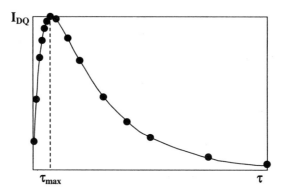

Figure 3 *Dependence of the peak height of double quantum filtered ^{23}Na spectrum (I_{DQ}) on creation time, τ. The solid curve is the fit to equation (2).*

3 RESULTS AND DISCUSSION

3.1 Pork meat

For both nuclei, and irrespective of salt concentration, ^{23}Na and ^{35}Cl SQ peaks were well defined by a single Lorentzian. The biexponential relaxation character of sodium and chloride ions (not resolved in SQ experiments), was systematically detected in DQF spectra. This indicates that sodium and chloride ions experience anisotropic or restricted motion (bound, entrapped ions). This means that a local order in pork samples exists at a microscopic scale.

The dependence on salt concentration of T_{2s}, T_{2f} and τ_{max} obtained from ^{23}Na T_2-DQF experiment is shown in Figure 4. Slight but significant variations in the DQ transverse relaxation times were observed. The τ_{max} dependence on salt content, which is directly correlated to those of T_{2s} and T_{2f} (Equation 3) demonstrates that the determination of τ_{max} is of crucial importance if exact DQF signal quantification is to be performed, and accurate comparison with the SQ signal made.

Taking into account this constraint for both sodium and chloride ions, DQF and SQ spectra were quantified. Summing the different contributions of "bound" and "free" ions thus determined, a satisfactory correlation with sodium chloride concentration expected in the pork meat samples studied was obtained ([NaCl]$_{NMR}$ = 1.01×[NaCl]$_{added}$; R^2 = 0.95).

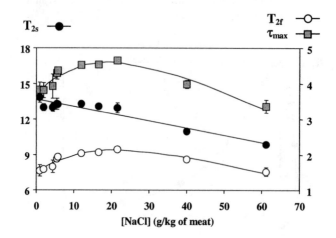

Figure 4   *Variations of T_{2s}, T_{2f} and τ_{max} determined from T_2-DQF experiments for Na$^+$ as a function of the sodium chloride content in minced pork. All times are expressed in ms.*

3.2   Smoked salmon

^{23}Na and ^{35}Cl single quantum spectra of smoked salmon unequivocally show (as displayed in Figure 5 for ^{35}Cl nucleus) the presence of a broad line superimposed on a narrow line. This is a clear indication that sodium and chloride ions in smoked salmon occur in more than one chemical state.

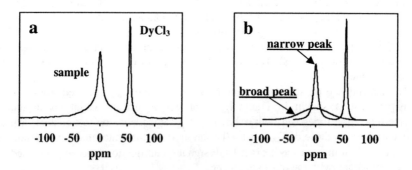

Figure 5   *^{35}Cl SQ spectrum of smoked salmon (a) and the result of the deconvolution (b).*

Sodium chloride quantification was performed from the analysis of these SQ spectra. Results are reported in Table 1 for the two technological processes A (dry salt) and B (dry salt and brined injection combination), and for the two extracted sample locations (thick (a) and thin (b) parts of the fillet). The two processes are clearly discriminated. The total salt content values obtained for samples b are approximately twice those of samples a. This marked heterogeneity of salt distribution may be qualitatively explained by a greater penetration of salt in the thin part than in the thick part of the fillet. Another interesting though surprising result is the relative contributions of the two ions, Na^+ and Cl^-, to the total salt content, about 1.6 to 1.8 times larger for sodium ion (taking into account the atomic mass of each ion). However, the total salt content values obtained, irrespective of process or sample location, are consistent with those expected. The development, in progress, of specific electrodes dedicated to sodium and chloride ion quantification in such products should help validate these results.

Table 1 *Sodium and chloride ions concentration in smoked salmon, determined by the analysis of SQ NMR spectra as described in Figure 5-b, for the two processes A (dry salt) and B (dry salt + brined injection), and samples a (thick part of the fillet) and b (thin part of the fillet)*

		Process A		**Process B**	
		Na^+	**Cl^-**	**Na^+**	**Cl^-**
Sample a	Broad peak	0.77	0.61	0.76	0.53
	Narrow peak	0.58	0.52	0.42	0.44
	Total ion	1.35	1.13	1.18	0.97
	Total salt	*2.48*		*2.15*	
Sample b	Broad peak	1.62	1.17	1.32	0.95
	Narrow peak	1.12	1.36	0.73	0.97
	Total ion	2.74	2.53	2.05	1.92
	Total salt	*5.27*		*3.97*	

values are expressed in g of ion per 100 g of smoked salmon.

4 CONCLUSION

The different NMR approaches described here demonstrate the wealth of information that can be obtained by ^{23}Na and ^{35}Cl spectroscopy for the characterisation of salted meat products. Parallel sensorial analyses are in progress to correlate NMR results with the perceived saltiness of these products. In particular, we need a more precise understanding of the respective roles of sodium and chloride ions in the salted taste, and especially the effects of the bound/free ratio of each ion.

References

1   G. Navon, H; Shinar, U. Eliav and Y. Seo, *NMR Biomed.*, 2001, **14**, 112.
2   P.S. Belton, K.J. Packer and T.E. Southon, *J. Sci. Food Agric.*, 1987, **41**, 267.
3   T. Nagata, Y. Chuda, X. Yan, M. Suzuki and K. Kawasaki, *J. Sci. Food Agric.*, 2000, **80**, 1151.
4   S.C. Chu, M.M. Pike, E.T Fossel, T.W. Smith, J.A. Balschi and S. Springer Jr., *J. Magn. Res.*, 1984, **56**, 33.
5   G. Jaccard, S. Wimperis and G. Bodenhausen, *J. Chem. Phys.*, 1986, **85**, 6282.
6   G. Bodenhausen, H. Kogler and R.R. Ernst, *J. Magn. Res.*, 1984, **58**, 370.
7   G.S. Payne, A-M.L. Seymour, P. Styles and K. Radda, *NMR Biomed.*, 1990, **3**, 139.

NMR STUDIES OF COMPLEX FOODS IN OFF-LINE AND ON-LINE SITUATIONS

B.P.Hills[1] , L. Meriodeau[2] and K.M.Wright[1]

[1]Institute of Food Research, Norwich Research Park, Colney, Norwich NR4 7UA, UK
[2] ENSBANA, Universite de Bourgogne, 1 Esplanade Erasme, 2100 Dijon, France.

1. INTRODUCTION

Most real foods are extremely complex microheterogeneous, multicomponent and multiphase biopolymer systems. Understanding how water partitions between the various components and microphases as foods are processed and stored remains a fundamental problem in food science. Indeed, a quantitative understanding of the factors controlling the microscopic redistribution of water is essential for predicting the effects of processing and storage on food quality. Finding experimental techniques to monitor the microscopic water redistribution also presents an outstanding challenge, and NMR relaxometry and diffusometry are undoubtedly the techniques of choice in this endeavour. In this chapter we therefore begin by reviewing the principles underlying moisture migration in microheterogeneous food biopolymer systems and show how NMR can be applied to the problem.

While such off-line laboratory NMR studies contribute to a fundamental understanding of the behaviour of food as it is processed and stored, there are additional complexities if we attempt to use NMR to monitor food "on-line" as it is sorted, processed and stored in a factory situation. Principle among these problems is the effect of magnetic field inhomogeneity on simple NMR acquisition sequences. This effect will be analysed and some of the outstanding challenges in transferring NMR technology to the on-line industrial situation will be discussed.

2. MOISTURE REDISTRIBUTION IN MULTICOMPONENT, MULTIPHASE BIOPOLYMER SYSTEMS

Figure 1 shows an optical micrograph of a representative multicomponent, multiphase, micro-heterogeneous food system which, in this example, consists of partially gelatinised heated pea starch granules embedded in an egg albumin protein matrix. Although the optical micrograph clearly reveals the microstructure, it fails to give any information about the distribution of the water among the major biopolymer components (amylose, amylopectin and egg albumin) or between the microphases. In fact higher resolution micrographs reveal that there are at least three distinct microphases. Besides the starch granule phase, the egg albumin phase has undergone phase separation into a microphase of phase-separated amylose and a concentrated albumin microphase. In this example, we expect that as the temperature is raised the starch granules swell and suck water out of the albumin phase until the gelatinisation temperature is reached. Thereafter, the release of amylose from the granules carries water back into the albumin phase. The amount of water migration determines the gelatinisation temperature and the

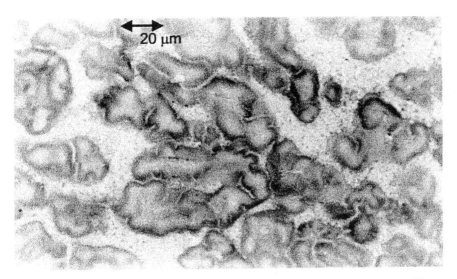

Figure 1. *Partially gelatinised pea starch granules in an egg albumin matrix. (Courtesy of Dr.M.Parker, Institute of Food Research, Norwich).*

rate of phase separation, so it is a prime factor in controlling the processing response in this system. Until this redistribution is understood we cannot predict the effects of modifying the structure of the starch granules or the effect of altering the composition of the albumin phase. Moisture migration between components and microphases during storage also affects shelf life by altering rates of starch retrogradation and phase separation and therefore, ultimately, the texture and microbial stability.

We can begin to develop a theory of the moisture redistribution by making the major assumption that the system is always in thermodynamic equilibrium. If so the water chemical potential, μ in all phases and components must be equal. Taking the textbook definition of the water potential, as $\mu = \mu_0 + RT\ln a_w$ in a non-ideal solution, this means that the water activity, a_w, must everywhere be the same. Knowledge of the sorption isotherms, $a_w(W_i)$ for each component, i, such as egg albumin or starch, should, therefore be sufficient to predict the equilibrium water content, W_i, in each component and microphase. Unfortunately, things are not so simple because it is very unlikely that the system really is in equilibrium. If one or more microphases enters the glassy state or becomes very concentrated then moisture diffusion coefficients may become so small as to be rate limiting. Interfacial barriers such as surface structures at the starch granule/albumin interface could also act as moisture permeability barriers. Slow macromolecule diffusion could also arrest microphase separation and moisture redistribution. All these factors mean that it is unsafe to assume thermodynamic equilibrium. Even if the system really is at equilibrium, it is naïve to assume that the water chemical potential in a microheterogeneous system is given by a formula describing the deviation of aqueous solutions from ideality. In heterogeneous systems there are other free energy terms to consider, such as surface free energies, swelling pressure terms and mixing contributions that are extremely difficult to evaluate *a-priori*.

These extra free energy terms mean that the water activity is no longer uniform, but varies on the microscopic distance scale of the system heterogeneity[1]. This point is often overlooked by food microbiologists who all to often rely on global water activities measured from the "equilibrium" vapour pressure, to determine the microbial shelf-life of a food. A typical food-borne bacterial pathogen has a size of 1-2 microns and therefore probes the water availability on this microscopic distance scale where microheterogeneity is all important and the "water activity", however we choose to define it, shows microscopic variation. We shall see some examples of this in a later section.

The difficulties in theoretically predicting water relations in multicomponent, multiphase systems means that it is all the more important to have experimental techniques capable of measuring microscopic water distribution. Band shape analysis of water peaks in FTIR may offer some hope in this regard[2], but the most promising technique is undoubtedly NMR relaxometry. Figure 2 shows the "transverse relaxation time spectrum" (i.e. the distribution of proton transverse relaxation times) measured with the CPMG pulse sequence for a suspension of potato starch granules in a 30% BSA solution as a function of thermal processing temperature. In this experiment, the starch suspension was heated to the indicated temperature for 20 minutes, then cooled to 23^0C, where it was measured with the CPMG pulse sequence with a 90-180 pulse spacing of 240μs. The resulting decay curve was analysed as a continuous distribution of exponentials using the standard Resonance Instruments DXP software. Each peak in figure 2 corresponds to water in a different microphase, though it is not always easy to assign them to specific structures. The peak at ca 50ms in the unheated sample is believed to correspond to water in the BSA solution outside the granules and the broad peaks between 1-10ms to water inside the starch granules[3,4]. Heating the sample at or above the gelatinisation temperature of ca. 60 degrees causes the release of amylose and water from the granule into the surrounding BSA phase and this can be quantified as an increase in the area of the peaks above ca 6ms.

Difficulties in peak assignment is one obvious shortcoming of this approach, as is the unknown extent to which water diffusion between different microphases averages the signal. The later can be minimised by working at temperatures just above freezing, and peak separation can be altered by modifying the pulse spacing and by varying the extent of diffusive attenuation by imposition of constant field gradients during the CPMG acquisition[6]. Alternatively, deuterium relaxometry in a D_2O exchanged system could give insight by removing the signal from non-exchanging biopolymer protons. Clearly much development work is still needed in multicomponent, multiphase water relaxometry and diffusometry before the technique can be used quantitatively for monitoring microscopic water redistribution.

Even if we assume the system is in thermodynamic equilibrium, we still need to understand the molecular and microscopic factors determining the water activity in multicomponent, multiphase systems. Such a theory needs to be able to predict the effect of biopolymer denaturation, aggregation and gelation on the water activity. The author has made a first, admittedly crude, attempt at such a theory in the final chapter of his book, "Magnetic Resonance Imaging in Food Science"[6]. The theory recognises the fact that water molecules hydrating a biopolymer have a range of mean exchange lifetimes from microseconds to picoseconds (see article by Prof. K.Wuthrich). While there is a continuous distribution of such water exchange lifetimes, we can simplify the situation by classifying the distribution into three distinct subpopulations of water, namely, free

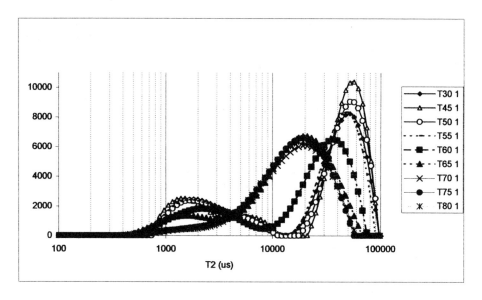

Figure 2. *Water proton transverse relaxation time distribution for potato starch in a 30% BSA solution that has been heated to the indicated temperatures. All measurements taken at 30⁰C*

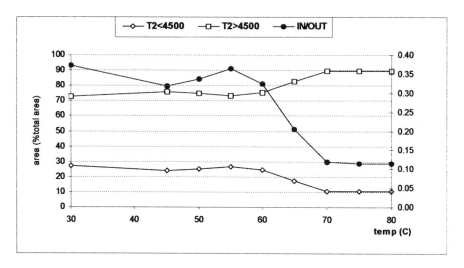

Figure 3. *The temperature dependence of the ratio of water inside/outside the starch granules deduced from the data in figure 2.*

bulk water, the mono-or bilayer of hydration water weakly hydrogen-bonding the surface of the biopolymer; and "structural water", which is water inside the biopolymer with relatively long exchange lifetimes and whose removal causes conformational change to the biopolymer. We further assume that each of these states can be assigned its own intrinsic water activity, a_i. This may not be such a bad assumption if we remember that each water state will differ in its free energy content. The Ergodic theorem can then be used to calculate the ensemble averaged water activity,

$$a_{av} = \lim_{t \to \infty} 1/t \int_{t0}^{t0+t} dt\, a(t) = \Sigma_i\, x_i a_i \qquad [1]$$

This describes a single component system. In a multicomponent system composed of biopolymers j; and water state i, it can be generalised to,

$$a_{av} = \Sigma_{ij}\, \xi(j)x_i(j)a_i(j) \quad \text{where } \Sigma_{ij}\, \xi(j)x_i(j) = 1 \qquad [2]$$

where $\xi(j)$ are preference coefficients. An analogous theory can be used for the water relaxation rates, R,

$$R_{av} = \Sigma_{ij}\, \xi(j)x_i(j)R_i(j) \quad \text{where } \Sigma_{ij}\, \xi(j)x_i(j) = 1 \qquad [3]$$

Which implies that R_{av} is linearly related to the average water activity:

$$R_{av} = R_{bulk} + \text{Const.}(1 - a_{av}) \qquad [4]$$

Where the constant depends on which states are present.
We can make further progress using a single correlation time model for the water dynamics in each state, i, such that, for transverse relaxation, and each component, j,

$$R_{2i} = (2c/3)[J(0,\tau_i) + (5/3)J(\omega,\tau_i) + (2/3)J(2\omega,\tau_i)] \qquad [5]$$

where the J's are the usual spectral densities,

$$J(n\omega,\tau_i) = \tau_i/(1+n^2\omega^2\tau_i^2) \qquad [6]$$

and τ_i is the motional correlation time of water in state i. Unfortunately, to extract the correlation times for each state, τ_i, from a measurement of R_{av} requires knowledge of the fraction of water in each state for each biopolymer, $x_i(j)$. High resolution NMR might provide this information in dilute solutions but not in the concentrated heterogeneous systems we are considering. Another possibility is to extract it by deconvoluting the spectral lineshape for the water peak in FTIR[2]. NMR field cycling relaxometry offers another possible tool for studying water dynamics and distribution in these complex systems, especially in low water content systems where the water dynamics is slowed to the dynamic window probed by the longitudinal spectral density function. Figure 3 shows an example where the water correlation time is so slowed by interaction with the sugar matrix in a sucrose glass that its spectral density can be extracted directly for the T_1 -frequency dispersion by straightforward deconvolution[7]. Clearly this type of measurement could be usefully applied to low water content

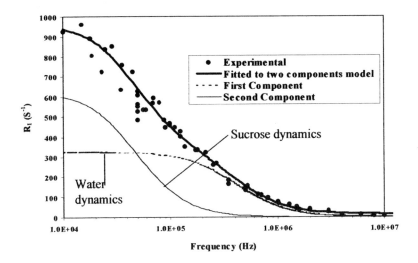

Figure 4. *The double T_1 frequency dispersion for a 95% sucrose solution at x K, deconvoluted into the spectral densities for water and the sucrose matrix. Taken from reference 7.*

multicomponent, multiphase biopolymer systems.

3. TRANSVERSE RELAXATION TIME SPECTRA AS PROBES OF FOOD SPOILAGE DURING STORAGE

The difficulty in interpreting proton transverse relaxation time distributions in complex foods does not prevent their use as empirical probes of food quality. Figure 5 shows an example where the relaxation spectrum shows distinct changes as a chocolate undergoes surface blooming. This blooming is believed to be caused by a polymorphic phase transition in the lipid phase driving a microscopic redistribution of water in the surface layers. NMR relaxation time spectra might therefore be useful early warning indicators of the onset of chocolate bloom. Other examples where changes in the relaxation spectra have been observed include the deterioration of toffees and wafers stored over long periods. Similar quality changes resulting from microscopic water redistribution are seen in fruit and vegetables. The development of "mealiness" in apples and peaches and tissue breakdown in melons are typical examples where the redistribution of water at a cellular and sub-cellular level results in changed mouth feel and undesirable loss of juiciness and texture. Not surprisingly these microscopic changes are clearly seen in the NMR water proton transverse relaxation time distributions and result in changes in the average transverse relaxation times and to image contrast changes in relaxation weighted NMR images of mealy fruit[8]. This suggests that relaxation weighted MRI images could be developed as a novel on-line sensor to sort fruit on a moving conveyor in a pack-house. This is indeed an exciting possibility but there are substantial technical hurdles to be overcome before this becomes practical and some of these difficulties are discussed in a later section.

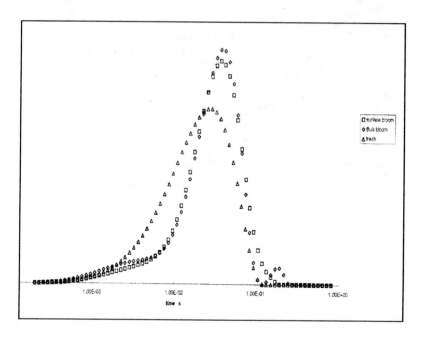

Figure 5. *Proton transverse relaxation time distributions for fresh and bloomed chocolate.*

4. TRANSVERSE RELAXATION TIME SPECTRA AND THE MICROBIAL STABILITY OF PROCESSED FOODS

Transverse water proton relaxation time spectra are also useful tools for predicting the microbial stability of microheterogeneous foods. The fact that the microscopic redistribution of water can strongly affect food microbial stability (and hence shelf life) is often overlooked by food microbiologists. Figure 6 shows a clear example of such a microstructural effect[9]. In this experiment, a synthetic bacterial growth medium(MOPS) was inoculated with 10^7 cfu/ml of *E. coli* and the plots show the decrease in the number of surviving, viable cells as a function of increasing incubation time. The experiment was repeated with three systems all having the same water activity, 0.94, (measured by the vapour pressure), temperature, pH and nutrient composition and differing only in the microscopic distribution of the water. In the top curve the water activity was controlled by addition of (non-metabolised) sucrose which dissolved and gave a homogeneous solution, lacking any microstructure. In the other plots the water activity was controlled by addition of Sephadex microspheres and fine silica granules, which caused a microscopic redistribution of the water. The effects of this microscopic redistribution can be seen to be quite dramatic. This microstructural stress effect can be usefully correlated with the amount of bulk water measured with the relaxation time distribution. Figure 7 shows this for the Sephadex system and the implications of this have been discussed elsewhere[10]. Most recently the effects of in-situ microstructure development on pathogen survival has been studied for a system undergoing spinodal decomposition[11], where equally dramatic changes are observed.

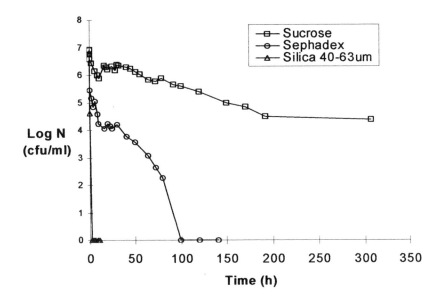

Figure 6. *The microstructural stress effect. Challenge test of E.coli K-12 (frag 1) in sucrose solution, Sephadex G25-50 and silica all poised at a water activity of 0.94, 200C. Taken from reference 9.*

Bearing in mind that most food microbiologists are not NMR specialists, these experiments suggest that we need to incorporate microstructural effects into existing models of microbial growth and survival as a simple number ranging from 0 to 1, much like water activity. Whether such a simple numerical "microstructural stress factor" can be meaningfully extracted from the NMR relaxation time distributions in combination with optical micrographs remains to be explored.

5. NMR RELAXOMETRY AS AN ON-LINE SENSOR OF FOOD QUALITY

The previous example where apple mealiness was correlated with the NMR water proton transverse relaxation time is only one of many such correlations that have been reported for many different food categories in the NMR research literature. Although such correlations emphasise the power of NMR in food science it also highlights a grave deficiency in the NMR research effort. Despite these numerous academic publications

NMR has had very little real impact on the food industry. This is especially surprising when we note that almost every other type of spectroscopy has given rise to commercially viable sensors on the production line. Near Infra-red (NIR) sensors, X-ray sensors, ultrasonic sensors and optical sensors, to name but a few, are routinely found on industrial production lines where they are used in a fully

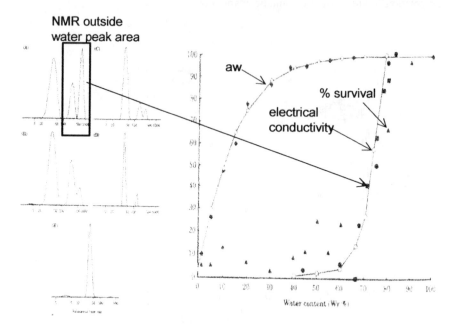

Figure 7. *Correlations between NMR measurements of external water and microbial survival in a packed bed of Sephadex microspheres containing a synthetic growth medium. Taken from reference 10.*

automated mode to detect foreign bodies and defective products. Yet, apart from a few rare examples in niche applications, there are no automated NMR or MRI sensors on industrial production lines. As an NMR specialist, this situation has always struck me as somewhat bizarre. It is true that simple, low-field bench-top NMR spectrometers are often used in an off-line mode in a quality control laboratory next door to the food processing plant to measure properties such as solid-liquid ratios and water content but this is a pitiful state of affairs for a mature technique that has been researched for over 50 years. There are thousands of research papers in the NMR literature showing useful correlations between NMR properties and the internal quality of materials as diverse as foods, wood, minerals and concrete, yet most of these correlations are unexploited in on-line quality assurance. In my opinion, transferring these correlations from the academic world to the factory floor represents one of the greatest challenges facing NMR researchers today. It is therefore worthwhile looking briefly at some of the technical hurdles that need to be overcome if "on-line NMR sensors of food quality" are ever to become a reality. The first challenge can be labelled "sample polarisation". In laboratory NMR we usually assume the sample has attained equilibrium longitudinal magnetisation before we start the NMR acquisition sequence, and the only situation where we need to be concerned with sample polarisation is when we use multiple acquisitions (accumulations) in which case recycle delays of at least $5T_1$ are needed if we are to avoid saturation effects. However, if the sample is travelling on a conveyor belt at a speed, v, the sample will have moved a distance $v5T_1$ in a magnetic field before it is fully polarised, and for typical conveyor velocities of 2m/s and a representative T_1 of about a second the sample has travelled 10 meters down the magnet before we even start the NMR acquisition! Obviously we can shorten this distance by working with sub-optimum polarisations which compromise the signal/noise, but even then, the polarisation time will limit the number of acquisitions that are possible. Unfortunately, stopping each sample in the sensor for the polarisation and acquisition is not a commercially viable possibility!

The high cost of commercial NMR spectrometers is another factor hindering their development as on-line sensors. Most modern systems are based around expensive superconducting magnets because they provide high field strengths and do not require a power source. However the high cost of these magnets means this technology is inappropriate for most factory situations. Permanent magnet systems are more robust and cheaper but the difficulties in shimming permanent magnet systems over the large volumes needed when a sample is moving at speed through the magnet are prohibitive. This leaves only resistive solenoid magnets, which were first used in the early pioneering days of NMR. The problems associated with resistive magnets are well known. The high currents needed for creating substantial magnetic field strengths deposit large amounts of heat in the coils, which necessitates expensive water-cooling. Special steps are also needed to avoid current fluctuations and temperature variations in the magnet.

Perhaps the most serious hurdle to be overcome in the development of on-line NMR sensors is the requirement for a magnetic field of sufficiently high spatial homogeneity for meaningful NMR. Most NMR researchers using sophisticated laboratory spectrometers take it for granted that, after shimming, there is field homogeneous to a few ppm over the sample volume. But when the sample is moving and multiple acquisitions are required, we are faced with the problem of shimming quite large volumes. This is technically possible and is done in whole body imagers, but the design of magnet and shim coils that permits this high degree of field homogeneity is an expensive and technically demanding exercise, and this hurdle has to be faced by anyone aspiring to design a commercially viable on-line NMR sensor. It is therefore worth analysing in some detail the effects of field inhomogeneity on the NMR of a discrete sample, such as an apple, moving at constant velocity through the field.

As a first step let us expand the inhomogeneous magnetic field at a point, r, as a Taylor series:

$$\mathbf{B}(r+\Delta r) = \mathbf{B}_0(r) + \mathbf{G}_{eff}(r)\Delta r + \ldots \qquad [7]$$

This shows that the field inhomogeneity can, to first order, be represented as an effective local gradient, $\mathbf{G}_{eff}(r)$, whose direction and magnitude will vary with position, r. Let us therefore analyse the effect of these local gradients on the FID of a rigid sample, such as an apple, moving at constant velocity, $\mathbf{v}$, through the sensor field. For the time being effects of spin relaxation, spin diffusion, and spin couplings are ignored. Consider spins in the volume element dV at position vector $\mathbf{r}(0)$ at time 0. After a time t, this element will have moved to $\mathbf{r}(t) = \mathbf{r}(0) + \mathbf{v}t$, where $\mathbf{v}$ is the linear velocity. For notational convenience we drop the subscript "effective" in the field gradient, $\mathbf{G_{eff}}$. The precession frequency at $\mathbf{r}(t)$ will then be

$$\omega(\mathbf{r}(t),t) = 2\pi f(\mathbf{r}(t),t) = \gamma B_0 + \gamma \mathbf{G}.\mathbf{r}(t)$$
$$= \gamma B_0 + \gamma \mathbf{G}.\mathbf{r}(0) + \gamma \mathbf{G}.\mathbf{v}t \tag{8}$$

As the spins initially at $\mathbf{r}(0)$ are carried along in the field gradient, they will accumulate a net phase angle given by

$$\varphi(\mathbf{r}(t),t) = \int_0^t dt'\, 2\pi f(\mathbf{r}(t'),t')$$
$$= \gamma B_0 t + \gamma \mathbf{G}.\mathbf{r}(0)t + \gamma \mathbf{G}.\mathbf{v}t^2/2 \tag{9}$$

The spin density $\rho(\mathbf{r}(t),t)$ of spins at $\mathbf{r}(t)$ at time t is clearly equal to $\rho(\mathbf{r}(0),0)$ if we ignore diffusion and bulk motion of the sample other than the linear translation, $\mathbf{v}$. The contribution $dS(\mathbf{r}(t),t)$ to the total signal $S(t)$ from spins at $\mathbf{r}(t)$ is therefore given by

$$dS(\mathbf{r}(t),t) = A.\ \rho(\mathbf{r}(t),t).\ \exp\left[i\varphi(\mathbf{r}(t),t)\right]$$
$$= A.\ \rho(\mathbf{r}(0),t).\exp[i\gamma B_0 t + i\gamma \mathbf{G}.\mathbf{r}(0)t + i\gamma \mathbf{G}.\mathbf{v}t^2/2] \tag{10}$$

The phase factor $\exp(i\gamma B_0 t)$ is eliminated by setting the RF coil "on-resonance" so, neglecting the constant of proportionality, A, and writing $\mathbf{r}(0)$ simply as r, and $\rho(\mathbf{r}(0),t)$ simply as $\rho(r)$ this becomes,

$$dS(\mathbf{r}(t),t) = \rho(r).\exp[i\gamma \mathbf{G(r)}.\mathbf{r}t + i\gamma \mathbf{G(r)}.\mathbf{v}t^2/2] \tag{11}$$

The total signal $S(t)$ is obtained by integrating over the whole sample volume,

$$S(t) = \int dV\ \rho(r).\exp[i\gamma \mathbf{G(r)}.\mathbf{r}t + i\gamma \mathbf{G(r)}.\mathbf{v}t^2/2] \tag{12}$$

Various limits of this expression can be considered. If the sample, such as an apple, is stationary and $\mathbf{G}$ is an applied linear imaging gradient, then the r-dependence of $G(r)$ can be dropped and the expression reduces to the conventional imaging algorithm

$$S(t) = \int dV\ \rho(r).\exp[i\gamma \mathbf{G}.\mathbf{r}t] \tag{13}$$

Fourier transforming the signal, $S(t)$ then gives the image, $\rho(r)$. But now suppose the sample is stationary and the field gradient, $\mathbf{G}(r)$, is not a linear applied imaging gradient, but rather represents a part of a non-linear field gradient resulting from inhomogeneity in the main field, B_0. In this case the exponential term will cause dephasing of the spins in different parts of a finite-sized sample such as an apple, simply because spins in different parts of the sample experience different local fields and field gradients. The effect will be reflected in a decay of the FID with a time constant, T_2^*, which contains no useful information about sample quality, merely the magnitude of the local field inhomogeneity. Of course, with a stationary sample this type of dephasing can be refocused in a spin echo, and this permits meaningful measurement of transverse sample relaxation times. However, if the field inhomogeneity is very severe then dephasing will be so rapid that T_2^* is short compared to the ring-down time of the RF probe so no FID will be observed, though it may still be possible to see a spin echo if a very short echo time is used.

Additional complications arise if the sample is travelling with velocity, **v**, through an inhomogeneous field. Not only is there additional dephasing from the term, $\gamma G_{eff}(\mathbf{r}).\mathbf{v}t^2/2$, where $G_{eff}(\mathbf{r})$ differs from point to point in the magnet, there is an even more serious problem arising from the sample motion because the dephasing cannot necessarily be refocused as a meaningful spin echo. This is because each moving spin will experience different local effective field gradients before and after the refocusing 180-degree pulse, so that it will no longer be returned to its initial phase at the expected echo time 2τ. Even if a spin-echo can be observed from a moving sample in an inhomogeneous field it will not necessarily contain any useful information about the nature of the sample, such as its transverse relaxation time, T_2, but only information about the local magnetic field inhomogeneities the spins have experienced in their journey during the echo time! This, no doubt, accounts for the observation that echo attenuation increases with increasing speed as a sample moves through a laboratory horizontal bore NMR spectrometer (ref).

Rapid dephasing of moving samples by field inhomogeneities also means that we require RF coils with short ring-down times if we are to have any hope of picking up NMR signals. Because the ring-down time is $2Q/\omega_0$, this means we must design RF probes with low Q factors. Low Q factors are also needed to avoid over-sensitive probe tuning whereby the tuning varies from one sample to the next, which is potentially disastrous in an on-line situation. There is therefore a compromise to be made between high signal/noise, requiring a high Q factor, and short probe ringdown times and low tuning sensitivity which requires a low Q factor.

These various considerations illustrate some of the major scientific and technical problems to be overcome if conventional laboratory NMR is to be adapted as a sensor of food quality in an on-line situation. Of course, all of the technical problems can be overcome at a price. Indeed, apart from the price tag, there is nothing preventing a whole body clinical imager being used in a pack house to monitor, for example, apple mealiness on a conveyor. However, even with a whole body scanner one would be restricted by sample speeds. An apple moving a 2m/s through the imager would not be fully polarised by the time it reached the central homogeneous part of the magnet and a very fast imaging sequence such as EPI or FLASH would be needed to acquire a 3D image before the sample departs. Unfortunately such rapid acquisition sequences require expensive and very carefully adjusted gradient controllers and coils. Indeed, just one standard gradient power amplifier costs about £10,000 and three would be needed for 3D imaging, which illustrates the need for developing new generations of low cost hardware for on-line applications. Few fruit or vegetable pack houses would be prepared to pay the prices of conventional NMR hardware for marginal improvements in sorting efficiency! All this points to the need for novel approaches to on-line NMR if commercially viable NMR sensors are to become commonplace on the factory floor.

Acknowledgements

The authors wish to thank the Biotechnology and Biological Science Research Council (BBSRC) for supporting this work from core Institute funding.

References

1. B.P.Hills and C.E.Manning, *J.Molec. Liq.*, 1998, **75,** 61.
2. J.Yarwood, C.Sammon, C.Mura, and M.Pereira, *J.Molec.Liquids*, 1999, **80**, 93.
3. H.Tang and B.P.Hills, *Carbohydrate polymers*, 2000, **43**, 375.
4. H.Tang and B.P.Hills, *Carbohydrate Polymers*, 2001, **46,** 7.
5. B.P., Hills, K.M.Wright and J.E.M. Snaar, *J.Magnetic Resonance Analysis*, 1996, **2,** 305.
6. B.P.Hills, Magnetic Resonance Imaging in Food Science, John Wiley & Sons, New York, 1998.
7. B.P.Hills, Y.L. Wang and H.R.Tang, *Molecular Physics*, 2001, **19,** 1679.
8. P. Barreiro, A. Moya, E. Correa, M. Ruiz-Altisent, M. Fernández-Valle, A. Peirs, K.M. Wright and B.P. Hills,. *Applied Magnetic Resonance*, 2002, **22,** 387.
9. B.P.Hills, L.Arnould, C.Bossu and Y.P.Ridge, *Int. J. Food Microbiology*, 2001, **66** 163.
10. B.P.Hills, C.E.Manning, Y.Ridge and T.Brocklehurst, *Int. J. Food Microbiology*, 1997, **36**, 187.
11. B.P.Hills, C.Buff, K.M.Wright, L.H.Sutcliffe and Y.Ridge, *Int. J. Food Microbiology*, 2001, **68,** 187.

DETERMINATION OF WATER SELF-DIFFUSION BEHAVIOR IN MICROORGANISMS BY USING PFG-NMR

C.H. Lee[1], Y.S. Hong[1], V.D. Volkova[2] and V. I. Volkov[3]

[1] Graduate School of Biotechnology, Korea University, 1 Anam-dong 5-ka, Sungbuk-gu Seoul 136-701, Korea. E-mail: chlee@korea.ac.kr
[2] Dept. of Foodstuffs, Moscow University of Consumer Co-operative, Voloshinej str. 12, Mytishi, Moscow region
[3] Laboratory of Membrane Processes, Center of Sciences and High Technologies, Karpov Institute of Physical Chemistry, Vorontsovo Pole 10, Moscow 103064, Russia. E-mail: vitwolf@mail.ru

1 INTRODUCTION

Water transport in biological systems is important for cellular physiological reactions, osmotic pressure of tissue and drying process of biological materials. Transport of water in cellular system occurs either due to osmotic gradients or due to molecular diffusion.[1] Osmotic water flow can be measured by monitoring changes in cellular light scattering[2] of concentration dependent fluorescence of cell size or of cell turgor pressure.[3] For diffusional water permeability, pulsed field gradient (PFG) NMR spectroscopy has become the method of choice due to its remarkable sensitivity to molecular displacements in the range of 10 nm-100 μm and to its non-invasive character.[4,5] It is not by chance, that the first pioneer PFG-NMR investigations of Tanner and Steiskal were carried out namely on biological cell.[6] Now PFG-NMR is widely applied to probe living tissue and biological cells structure, for measuring thermodynamic binding constants, membrane permeability and rates of transmembrane exchange processes.

Recently, Benga et al.[7,8] measured the diffusional water permeability of red blood cells from different species of animals including man by using PFG NMR, and Waldek et.al.[9] studied the effects of cholesterol on transmembrane water diffusion in human erythrocytes by the same method. Self-diffusion of water and oil in peanuts[10], and diffusion coefficient of bound water in cotton fiber and plant tissue[11] were measured by spin-echo nuclear magnetic resonance technique. Schoberth et al.[12] measured the diffusional water permeability in *Corynebacterium glutamicum* by using home-built PFG NMR spectrometer at a proton resonance frequency of 400 MHz with maximum field gradient amplitude of 24 T/m. With this equipment, the observation time was reduced to less than 1ms, which enables the measurement of self-diffusion coefficient of water in a compartment as small as a bacterial cell.

We investigated the self-diffusion coefficients of water in chlorella cells as a model of porous biological membrane and the changes in self-diffusion coefficients of yeast cells in different growth phases using home-built PFG NMR. The diffusional water permeability and average cell size were estimated from the self-diffusion measurements, and compared with those of measured by microscopic analysis.

2 MATERIAlS AND METHODS

2. 1. PFG NMR Measurement

The self-diffusion coefficient measurements were carried out with home-built PFG NMR machine, which was constructed at Center for Advanced Food Science and Technology (CAFST), Graduate School of Biotechnology, Korea University, Seoul, in collaboration with Prof. Vladimir Skirda from Department of Molecular Physics, Kazan State University, Kazan, Russia. The NMR frequency for protons was 63 MHz, and the maximum field gradient amplitude was 50T/m. The measurements were all computerized. The diffusion time (t_d), the interval between magnetic field gradient pulses, could be varied from 2 ms to 2 s. The duration time (δ) of magnetic field gradient pulse varied from 20 µs to 5 ms. The temperature of measurement varied from 20°C to 200°C with stability within ±0.5°C. The signal of free induction decay (FID) was converted by Fourier Transformation program installed in computer.

The details of PFG NMR had been described in many literatures.[14-17] In our measurement we used stimulated spin echo sequence with the magnetic field gradient pulses. This sequence is shown in Figure 1. During the measurement of echo signal amplitude evolution, τ and τ_1 are fixed, and only the dependence of A on g is analyzed, which is called the diffusional decay. The diffusional decay is expressed by the following equation.

$$A(g) = \frac{A(2\tau, \tau_1, g)}{A(2\tau, \tau_1, 0)} = \exp(-\gamma^2 \cdot g^2 \cdot \delta \cdot t_d \cdot D_s) \tag{1}$$

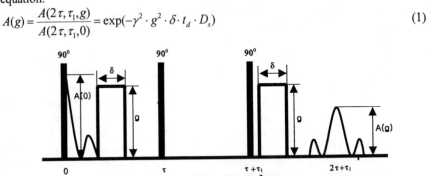

Figure 1 *Stimulated echo pulse sequence with the magnetic field gradient pulses. Here τ is the time interval between the first and second RF pulses, τ_1 is the time interval between the second and the third ones. Δ is the interval between the gradient pulses, δ is duration of the magnetic field gradient pulses, g is the amplitude of the gradient pulse. The gradient pulse is rectangular and oriented along the Z axis.*

Where γ is gyromagnetic or magnetogyric ratio, $t_d = \Delta - \delta/3$ is the diffusion time, and D_s is the self-diffusion coefficient, τ, τ_1 and g are shown in Figure 1. For our PFG NMR machine, the maximum values of g and δ are 50 T/m and 5×10^{-3}s, respectively. It gives us the opportunities to measure the diffusion coefficient from 10^{-8} to 10^{-15} m²/s and to observe diffusional decay changing in three orders of magnitude. The latter is very important when this decay is non-exponential.

In the case of non-exponential decays, the experimental curve A(g) are usually decomposed on the exponential components, which are described by equation (1). For the multiphase system consists of m phases in the case of slow (compare to t_d) molecular exchange between phases,

$$A(g) = \sum_{i=1}^{m} p_i \cdot \exp(-\gamma^2 \cdot g^2 \cdot \delta^2 \cdot t_d \cdot D_{si}), \qquad \sum_{i=1}^{m} p_i = 1 \qquad (2)$$

where D_{si} is the self-diffusion coefficient of i-th component and p_i is the relative amounts of nuclei belong to the molecules characterized by the self-diffusion coefficient D_{si}. The value p_i is usually called the population of i-th phase. By using the t_d dependence of self-diffusion coeffeicient, the permeabilty P, and the restricted area (cell size) a, were estimated for pore system by the following relationship.

$$\frac{1}{D_p} = \frac{1}{D_0} + \frac{1}{P \cdot a} \qquad (3)$$

where D_0 is non-restricted self-diffusion coefficient at $t_d \rightarrow 0$, D_p is hindered self-diffusion coefficient, where the total averaging of intrapore diffusion is achieved due to pore walls permeability.[16,17] The restricted size a can be determined from the plot of $D_s(t_d)$ the experimental self-diffusion coefficient dependence, where $D_s(t_d)$ is proportional to t_d^{-1}, if pore walls are not permeable. In the case of permeable pores, the restriction size and permeability maybe estimated from $D_s(t_d)$ dependences using sealing approach.[13-15] In this approach, the calculated dependences $D_s^{eff}(t_d)$ proportional to t_d^{-1} were analyzed, where

$$D_s^{eff} = \frac{[D_s(t_d) - D_p] \cdot D_0}{D_0 - D(t_d)} \qquad (4)$$

From $D^{eff}(t_d)$ slope, which is proportional to t_d^{-1}, the restricted size a was determined according to Einstein equation. The permeability, P^d was calculated from D_p value according to equation (3). In the estimation of cell size from t_d dependency of D_s, it was assumed that the water molecules in the cell move freely and the hindrance of intracellular particles against water movement was disregarded.

In order to calculate water molecules' residence time in the cell and the exchange rate constants, two phases model supposing exponential residence time distribution function was applied. From this model, the population of phase with the lowest self-diffusion coefficient D_{s1} depends on diffusion time t_d[4]

$$p(t_d) = p(0) \cdot \exp(-\frac{t_d}{\tau_r}) \qquad (5)$$

Where τ_r is water molecules' residence time in the cell.
The diffusional permeability, P^{eff} was estimated by exchange rate constant

$$P^{eff} = k \frac{V}{S} \qquad (6)$$

Where k is the molecular exchange rate constant, and

$$k = \frac{1}{\tau_r} \qquad (7)$$

The permeability, P_1^{eff}, was calculated from equation (6) by using the volume to surface area ratio, V/S from electron microscopic data.

2.2 Sample Preparation

Chlorella sp. (KCTC No. AG 10002) was obtained from Korea Research Institute of Bioscience and Biotechnology, Daejun, Korea. It was cultivated in a fermentor with organic medium.[18] The harvested cells were centrifuged to obtain chlorella pellet. It was washed three times in distilled water, and then filled in 5mm (external diameter) NMR tube (Series 300, Aldrich Chemical Co.). The dry matter of the cell pellet was 125 mg/g, which is equivalent to 87.5 % moisture content.

The yeast strain was *Saccharomyces cerevisea* isolated from Korean rice-beer, *Takju.*[19] The activated yeast was grown in YM broth (Difco, USA) in an shaking incubator at 27°C. The growth curve of yeast was obtained by measuring the optical density at 600 nm of the cultivation broth. A typical sigmoid type curve was obtained, consisting of induction lag phase, exponential growth phase and stationary growth phase. Three types of yeast cells were investigated, harvested at 9 hrs of incubation (mid exponential growth phase, MGP), at 24 hrs (end of exponential growth phase, EGP), and 48 hrs (stationary growth phase, SGP), respectively. The harvested cells were centrifuged to obtain yeast pellet. It was washed three times with phosphate buffer and filled in NMR tube. The weight dry matter of pellet was 192±3.4 mg/g, equivalent to 80.8 % moisture content.

2. 3 Electron Microscopy

The microstructure of the cell was investigated by TEM (H-600, Hitachi, Japan), and the cell topology was observed by SEM (Hitachi, S-450, Japan), which was set at an accelerating voltage of 15KV.[18] The cell pellets were washed several times with 0.1 M phosphate buffer (pH 7.0) and fixed in 2.5% glutaraldehyde at 4 °C and postfixed in 1 % OsO_4 for 2 hrs at room temperature. After washing again with 0.1 M phosphate buffer (pH 7.0), the specimens were dehydrated in graded ethanol (50%, 70%, 80%, 90%, 95% and absolute).

3. RESULTS AND DISCUSSION

The typical diffusional decay of water molecules in chlorella is shown in Figure 2. This decay was decomposed on three exponential components according to equation (2), which were characterized, by self-diffusion coefficients D_{s1}, D_{s2}, D_{s3} and populations p_1, p_2, p_3. D_{s3} was 2.7×10^{-9} m^2/s that is equal to bulk water self-diffusion coefficient at 30 °C.

The analysis of diffusional decay curves at different diffusion time t_d shows that the value of populations and self-diffusion coefficients D_{s1} and D_{s2} depend on t_d (Figure 3). D_{s1} decreased from 1.5×10^{-10} m^2/s at t_d 5ms to 7.5×10^{-12} m^2/s at t_d 200 ms, and the population decreased from 28% to 12%. Figure 4 shows that the self-diffusion coefficients D_{s1} and D_{s2} decrease with increasing t_d, but there was no dependence on t_d for self-diffusion coefficient D_{s3}, which was the self-diffusion coefficient of bulk water. The population p_1 and p_2 also decreased, but population p_3 increased with increasing diffusion time. The dependence of water self-diffusion coefficients on diffusion time indicates that water diffusion in chlorella is restricted, and D_{s1} and D_{s2} appear to represent the intracellular and extracellular water self-diffusion coefficients, respectively.[13] From these data the chlorella cell size (a), water molecule's residence time (τ_r), and exchange rate constant (k) were estimated and the cell wall permeability (P^d) was calculated.

Table 1 *The value of compartment size and permeability of chlorella pellet obtained from PFG-NMR data.*

	a (μm)	τ_r (ms)	k (s^{-1})	P^d (m/s)	P^{eff} (m/s)
Chlorella cell	3.3±0.3	220±20	4.5±0.5	$(6.0±1.2) \times 10^{-7}$	$(3.0±0.6) \times 10^{-6}$
Extracellular Cluster	17.0±3.0	320±60	3.1±0.6	$(2.8±0.6) \times 10^{-5}$	$(0.9±0.3) \times 10^{-5}$

P^d: permeabilities estimated by the t_d dependence of diffusion coefficient.
P^{eff}: permeability calculated by population data and V/S from electron microscopic data.

Table 1 summarizes the results. The estimated cell size, 3.3±0.3 μm, was slightly smaller than the size measured by TEM, about 5 μm, as shown in the photograph A of Figure 5. The discrepancies may be resulted by the fact that microscopic view shows the outer size of cell while the estimated *a* represents the inside distance. Therefore, the difference may indicate the cell wall thickness. The permeabilities estimated by the t_d dependence of diffusion coefficient, P^d, was 2-3 times bigger than those calculated by using V/S from electron microscopic data, P^{eff}. Photograph B in Figure 5 shows the cluster of chlorella cells in the aqueous suspension. The size of cell cluster estimated from D_s data was about 5 times bigger than that of individual cell, but the cluster size in the picture was about 30 times or more of the cell size.

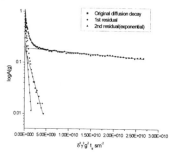 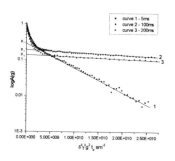

Figure 2 *The diffusion decay of water molecules in chlorella at t_d=100ms, δ=201μs, g_{max}=9.65T/m. Measurement temperature t=30 ℃.*

Figure 3 *The diffusion decays of water molecules in chlorella obtained at different t_d :1-5 ms (δ=900 μs, g_{max}=9.65T/m), 2-100 ms (δ=201μs, g_{max}=9.65T/m), 3-200 ms(δ=250 μs, g_{max}=9.65T/m). The temperature of measurements is 30 ℃.*

It may be explained by that, the behaviour of extracellular water, which involves in exchange processes with intracellular water and bulk water, is complicated compare to molecular exchange between two phases. For this reason, we can estimate the size and permeability of cell cluster by approximation only. It should be also mentioned that restriction size and permeability for clusters are the extracellular water restriction area and possibility of water molecues to connect to each other inside cluster respectively.

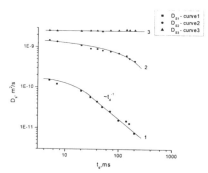

Figure 4 *The experimental dependence of self-diffusion coefficients D_{s1}, D_{s2}, D_{s3} on the diffusion time t_d of water molecules in chlorella sp. The temperature of measurements is 30 ℃.*

Similar diffusional decay curves were observed in yeast cells (Figure 6).[14] The analysis of diffusional decay curves at different diffusion time t_d showed that the values of populations p_1, p_2 and self-diffusion coefficient D_{s1} and D_{s2} were depended on t_d. Figure 6 shows that the self-diffusion coefficient of intracellular water varies with the cultivation time of yeast. Significant difference in D_{s1} was noticed between those of exponential growth phase and stationary growth phase.

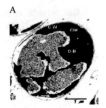 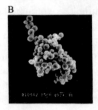

A   B

Figure 5 *Photography of chlorella cell cultured in fermenter by TEM and SEM. Scale bar corresponds to 1 μm for TEM (A) and 9.7 μm for SEM (B). C.W : Cell Wall, Chl : Chloroplast, D.B : Dense Body, St : Starch.*

On the other hand, D_{s2} was not affected by the age of yeast. The curve $D_{s1}(t_d)$ and $D_{s2}(t_d)$ are similar to ideal curve (Figure 6).[16,17] The slopes of these curves are less compared to t_d^{-1} (curves 1-3 and solid line in Figure 6). It may be concluded that the slopes $D_{s1}(t_d)$ decrease with increasing t_d, because of permeability effect.

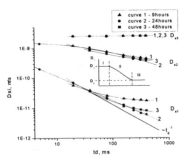

Figure 6 *The dependences of water self-diffusion coefficients D_{s1}, D_{s2} and D_{s3} on diffusion time t_d for the yeast harvested after (1) 9 hours incubation (MGP), (2) 24 hours (EGP) and (3) 48 hours (SGP). The temperature of measurement was 30°C.*

Table 2 summarizes the changes in cell size, water molecules residence time, exchange rate constant and cell wall permeability of yeast during cultivation. The estimated cell size increased from 2.3 μm in the middle of exponential growth phase to 3.0μm at the the end of exponential growth phase. The permeability decreased from 6.3×10^{-6} m/s to 8.4×10^{-7} m/s for the same period. The cell wall permeability decreased 6-7 folds by maturation of the cells from middle of exponential growth phase to the end of exponential growth phase. The slight increase in permeability during longer cultivation time (48hrs) in stationary growth phase may indicate cell wall aging and lyses. The residence time of water in the cell increased as the cells matured.

Table 2 *Cell size and permeability by growth phase of yeast as determined by PFG-NMR.*

	a (μm)	τ_r (ms)	k (s^{-1})	P_1^d (m/s)	P_1^{eff} (m/s)
MGP (9 hrs)	2.3±0.2	240±25	4.2±0.4	$(6.3\pm0.6)\times10^{-6}$	$(7.0\pm0.7)\times10^{-6}$
EGP (24 hrs)	3.0±0.2	450±40	2.2±0.2	$(8.4\pm0.8)\times10^{-7}$	$(1.2\pm1.0)\times10^{-6}$
SGP (48 hrs)	2.7±0.2	400±40	2.5±0.3	$(1.5\pm0.2)\times10^{-6}$	$(1.6\pm0.2)\times10^{-6}$

4. CONCLUSION

The present study demonstrates the applicability of PFG NMR for the measurement of water movement in and around cellular system. The cell wall permeabilities and cell size estimated from the diffusion time dependence of self-diffusion coefficient of water in restricted region agreed well with those obtained with microscopic data. The technique can be used for determining the changes in physiological conditions of living cells during growth and environmental changes. Further studies are needed to develop better measurement and knowledge in data interpretation for biological systems. More detail of present calculation methods and theoretical discussions are given in our recent publications.[13,14]

References

1   M.L. Garcia-Martin, P. Ballesteros and S. Cerdani, *Prog. Nucl. Mag. Res. Spect.*, 2001, **39**, 41.
2   P. Agre, J.C. Mathai, B.L. Smith and G.M. Preston, *Methods Enzymol.*, 1999, **294**, 550.
3   T. Henzler and E. Steudle, *J. Exp. Bot.*, 1995, **46**,199.
4   J. Karger, H. Pheifer and W. Heink, 'Principles and application of self-diffusion measurements by nuclear magnetic resonance' in *Advances in Magnetic Resonance*, ed. J.S. Waugh, Academic Press, New York, 1988, Vol. 12, p. 1.
5   Waldec, W. Kuchel, J. Lennon and E. Chapman, *Progr. Nucl. Magn. Reson. Spectr.*, 1997, **30**, 39.
6   J.E. Tanner and E.O. Stejskal, *J. Chem. Phys.*, 1968, **19**,1768.
7   G. Benga, S.M. Grieve, B.E. Chapman, C.H. Gallagher and P.W. Kuchel, *Comp. Haemat. Int.*, 1999, **9**, 43.
8   G. Benga, P.W. Kuchel, B.E. Chapman, G.C. Cox, I. Ghiran and C.H. Gallagher, *Comp. Haemat. Int.*, 2000, **10**, 1.
9   R. Waldek, M.H. Nouri-Sorkhbi, D.R. Sullivan and P.W. Kuchel, *Biophy. Chem.*, 1995, **55**, 197.
10 N.L. Zakhartchenko, V.D. Skirda and R.R. Valiullin, *Magn. Res. Imaging*, 1998, **16** 583.
11 A.V. Anisimov, N.Y. Sorokina and N.R. Dautova, *Magn. Res. Imaging*, 1998, **16** 568.
12 S.M. Schoberth, N.K. Bör, R. Krömer and J. Kärger, *Anal. Biochem.*, 2000, **279**, 100.
13 C.H. Cho, K. Kang, V.I. Volkov, V. Skirda and Cherl-Ho Lee, 2002 (submitted to *Mag. Res. Imaging).*
14 K.J. Suh, V. Skirda, V.I. Volkov, C.Y.J. Lee and C.H. Lee, 2002 (submitted to *Biophy. Chem.).*
15 Valiullin R. and V.D. Skirda, *J. Chem. Phys.*, 2001, **114**, 452.
16 P.P. Mitra, P.N. Sen and L.M. Schawartz, *Phys. Rev. B*, 1993, **47**, 8565.
17 P.P. Mitra, P.N. Sen, L.M. Schwartz and P.L. Doussal, *Phys. Rev. Lett.*, 1992, **68** 3555.
18 K. Kang, C.Y.J. Lee and C.H. Lee, *J. Microbiol. Biotechnol.*, 2002 (in press).
19 S.W. Yoon, C.Y.J. Lee, K.M. Kim and C.H. Lee *J. Microbiol. Biotechnol.*, 2002, **12**, 183.

ACKNOWLEDGEMENT

This study was partially supported by Korea Institute of Science and Technology Evaluation and Planning (KISTEP), Republic of Korea and Russian Foundation of Basic Researches, Grant No. 00-03-32099.

Food Quality Control

NMR RELAXOMETRY AND MRI FOR FOOD QUALITY CONTROL : APPLICATION TO DAIRY PRODUCTS AND PROCESSES

F. Mariette

CEMAGREF, Food Engineering research unit
17 avenue de Cucillé, CS64427, 35044 Rennes Cedex, France

1 INTRODUCTION

Since the first NMR studies concerning determination of the water content in milk powder[1,2] and determination of the solid fat index, relatively few new applications have been proposed for dairy companies. This situation could be explained by the difficulty of interpreting relaxation time parameters experienced for the non NMR specialist. On the other hand, the use of NMR for research in the dairy field is expanding yearly. These studies have involved the use of relaxation parameters to monitor the structural changes in the casein micelle and the consequences on water retention, the use of NMR diffusometry to study the effects of the dairy gel structure on the water self-diffusion coefficient and the use of MRI techniques to quantify the macroscopic organisation of dairy products. We present here some examples of the NMR and MRI results in the study of molecular and macroscopic changes in dairy products and how the NMR data can be used to provide better quality control throughout the different processing steps.

2 WATER RELAXATION MECHANISMS IN DAIRY PRODUCTS

Milk is a very fascinating natural product because of the complexity of the composition and structure. Milk can be described as an oil-in-water emulsion with fat globules dispersed in the continuous serum phase, a colloid suspension of casein micelles and a solution of lactose, soluble proteins and minerals. Although proteins represent only 3% of milk composition, casein (which represents 80% of the total amount of protein) plays a major role in the functional properties of milk. Caseins are organised as "micelles" which are large, roughly spherical aggregates, highly hydrated with about 4 – 6 g water/g protein[3,4]. These micelles are the major factor in water relaxation behaviour.

The NMR relaxation signal in milk is generally described by a mono-exponential equation, both for transverse and longitudinal relaxation decay curves. Multinuclear NMR studies[5] have demonstrated that the relaxation observed on skimmed milk shows that the relaxation mechanisms can be interpreted according to the self-diffusion exchange model proposed by Fedotov[6]. This general model considers three exchanging fractions of protons: free water protons (a), hydrating water protons (b) and exchangeable protons of the protein and lactose molecules (c). At low field NMR, the transverse relaxation rate is given by

$$\frac{1}{T_{2obs}} = \frac{P_a}{T_{2a}} + \frac{P_b}{T_{2b}} + \frac{P_c}{T_{2c} + k_c^{-1}} \quad [1]$$

where T_{2i} is the relaxation time of different proton states, with P_i their relative population, and k_c is the rate of proton exchange. Quantitative analysis of the relaxation value revealed that 58% of the proton relaxation rate came from hydrating water protons and 34% from the proton exchange mechanisms in the skimmed milk at pH 6.6 and 9% of dry matter[5]. Moreover it has been demonstrated that exchangeable protons come mainly from soluble whey proteins and lactose present in the water phase[5;7]. The hydrating water proton can be explained by the porous structure of the casein micelle which contains a large amount of water. This water is still highly mobile in fast exchange with the bulk water. The water relaxation is therefore the sum of the contribution of exchangeable proton from soluble proteins and water hydration embedded in the micelle. Because of the sensitivity of the water relaxation to casein micelle hydration, any physico-chemical treatments will modify the structure of the micelle and therefore the hydration properties can be monitored through relaxation time measurements.

3 WATER RELAXATION : A CASEIN STRUCTURE PROBE

Control of casein structure is one of the challenges for dairy product manufacturers to obtain the texture of final products. Integrity of the micelle involves a wide range of interactions such as hydrophobic bonding, hydrogen bonding, electrostatic interaction, disulphide bonding and calcium bonding. The main two industrial methods for destabilising the casein micelle structure are changes in pH, and addition of a proteolytic enzyme such as rennet. Although both induce formation of a gel, the destabilisation mechanism involved and the end characteristics of the gel are different according to the process used. The effect on the water relaxation will therefore also be different. When the pH is decreased from the natural pH of 6.6 to a value around 5.3, the micellar calcium and phosphorus (MCP) content decrease and induce considerable changes in the micellar voluminosity[8].The amount of water inside the micelle then decreases and the overall relaxation rate of water decreases independently of the temperature (Figure 1). Above pH 5.3 the concentration of the micellar MCP is nil, and the gel starts to be formed. At this pH range, the relaxation time of water reflects the change in gel structure induced by the temperature used during the acidification kinetics[9;10].

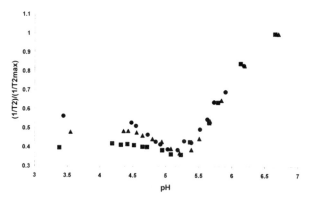

Figure 1 *Variations in normalised relaxation rate $1/T_2$ of reconstituted skimmed milk at 20 MHz with pH at different temperatures:* ● *5°C,* ▲ *20°C,* ■ *30°C.*

The advantage of the NMR technique to monitor pH changes is that the method can be used for whole milk and skimmed milk[11] and allows study of the decreased pH kinetic effect on micellar destabilisation[10]. Moreover, relaxation parameters are not sensitive to gel formation with rennet, and therefore structural micellar changes induced by lactic fermentation after addition of rennet can be selectively followed. Many experimental conditions have been studied (e.g. high pressure treatment of the milk[12]). In the latter case, variations in the relaxation time were measured according to pH for milk samples treated with different high pressure intensities. Between pH 6.8 and 5.2, changes in transverse relaxation time were enhanced by pressure treatment and demonstrated that the amount of micellar calcium phosphate was lower than in untreated milk.

The study on milk was also extended to concentrated casein samples. When the casein concentration increased, the variation in relaxation time to change in pH decreased (Figure 2). Two explanations can be considered. First, when the casein concentration increases, the buffer power of the sample increases and the MCP released is lower for low pH value[13]. Consequently, the pH effect on the casein structure is lower and little variation in T_2 is registered. Secondly, the sensitivity of the relaxation is mainly explained by the large difference between the relaxation time of water inside the micelle and the relaxation time of bulk water. For concentrated casein dispersion the amount of water outside the micelle decreases in relation to the water inside the micelle. The difference between the water relaxation in the two compartments then becomes small and the water relaxation time thus becomes less sensitive to acidification

If the pH effect can be monitored when it decreases, the water relaxation time can also be used to monitor increased pH. This has been well illustrated in experiments involving the ripening of soft cheese[14]. After enzymatic coagulation of the milk, the gel is drained and salted and then all the biochemical changes that occur during ripening begin. The lactate produced by the lactic bacteria during the acidification period is oxidized during ripening [15]. This metabolism of lactate causes an increase in pH. The combined action of pH and salt precipitation (especially calcium) changes the protein gel structure and softens the cheese.

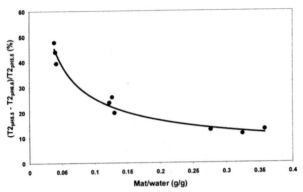

Figure 2 *Variations between the relaxation time T2 for casein retentate at pH 6.6 and pH 5.5 according to the casein concentration. T2 measurements were performed with a 20 MHz spectrometer (Bruker) at 37°C.*

These modifications in turn induce an increase in the water relaxation time. For example, the transverse relaxation time of the water phase was 35 ms for a commercial non-mature soft cheese and reached 60 ms when the cheese was mature. (Figure n°3).

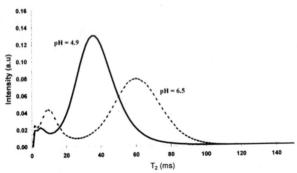

Figure 3 T_2 *Distribution from CPMG signal. Samples were soft cheeses, one non-ripened with pH=4.9 and the other ripened with pH=6.5 The first peak could be attributed to the liquid fraction of the fat, and the larger peak to the water phase. Measurements were performed at 5°C at 20Mhz.*

This variation in relaxation depends on the dry matter in the cheese: The lower the level of dry matter, the greater the variation. However, the sensitivity of the relaxation time to ripening is enough to use the relaxation time as parameters for application of MRI. Figure 4 represents the T_2 map of soft cheeses according to time of ripening. The T_2 map was calculated from a relaxation decay curve measured with a 2D Fourier transform multi-echo sequence[14]. After 9 days the relaxation time of the cheese was low and little heterogeneity was observed. After 14 days an increase in T_2 was observed on the periphery of the cheese. Moreover, there was a gradient from the centre to the cheese surface. The high T_2 value voxel number increased over time to the detriment of the low T_2 value voxel number. After 44 days, all the voxels reached the same T_2 values and the cheese was totally mature. The variation in T_2 observed during ripening was in close agreement with the expected ripening mechanism. It is known that deacidification occurs initially at the surface, resulting in a pH gradient from the surface to the centre of the cheese[15]. The MRI technique could therefore be applied to different ripening conditions, and the ripening kinetics could be followed. For example, an increase in the protein/water ratio slows down the ripening kinetics (Figure 5). The advantage of the MRI technique for application to soft cheese ripening is that morphological information such as the number and size of the opening[14;16;17;18;19;20], cheese volume, surface of the ring, and surface covered by the penicillium[14] could be easily quantified. Furthermore, the use of texture image analysis to quantify MRI cheese heterogeneity has also been evaluated. The textural features thus computerised allow discrimination of commercial soft cheeses[20] and the prediction of sensory characteristics such as sensory texture[21].

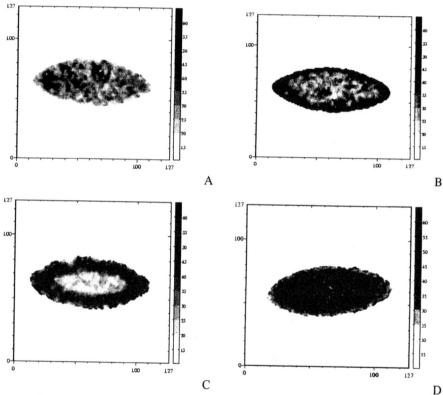

Figure 4 T_2 *map of soft cheeses according to ripening time. D + 9 (A), D+14 (B), D+30 (C) and D+44 (D). Images were acquired with an Open scanner (Siemens) at 0.2 T. A 24 multi-echo sequence was used with TE=17 ms. Temperature was controlled at 5°C.*

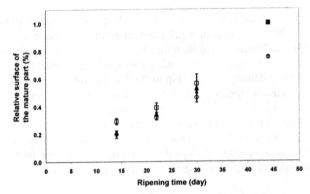

Figure 5 *Variation in relative surface area of the mature part of the cheese. The area was obtained after the threshold of the MRI T_2 map. The protein/water ratio was 0.33 g/g (O), 0.23 g/g($\square$) and 0.25 g/g ($\blacktriangle$).*

The use of water NMR relaxation time to quantify modifications of casein micelle structure is not limited to processing involving changes in pH. For example, relaxation time measurement has been used to study the effect of added salt on the hydration behaviour of casein micelle powder[22]. Another example involved the effects of protein - fat interactions on the protein structure in emulsion. It has been established that the extent of the structural changes induced by adsorption at the interface varies according to the native structure of the protein. According to Dalgleish[23], proteins can be divided into two structural classes, namely those which are essentially unstructured and flexible, such as casein, and those which possess well-defined three-dimensional structures such as globular protein. The adsorption to the oil-water interface causes denaturation of the latter type.

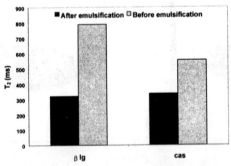

Figure 6 *T_2 water relaxation time in protein solutions before emulsification (grey bars) and in oil-water emulsion after emulsification (black bars). Measurements were performed at 30°C.*

This behaviour could be detected from relaxation measurements. For example, the variation in water relaxation time induced by formation of the emulsion was greater for β lactoglobulin than sodium caseinate (Figure 6). The emulsions were prepared from water solutions of 4% (w/v) β-lactoglobulin or sodium caseinate mixed with a vegetable oil (Mygliol + cocoa butter). The oil was emulsified (45 v/v) with a laboratory scale high pressure homogenizer (50Mpa) at 60°C. After emulsification the water relaxation time decreased by 39% for caseinate solution and 59% for β lactoglobulin solution. The decrease in T_2 relaxation observed for β lactoglobulin solution could be related to protein denaturation as observed after thermal treatment (Lambelet and al. 1992). A similar range of variations was measured for the T_2 relaxation ($\approx 65\%$).

4 WATER RETENTION, DISTRIBUTION AND DYNAMICS

4.1 Water retention and distribution

Water-holding properties have been recognised by food technologists among the many functional properties attributed to milk protein products. In general, water-holding is accomplished by complex interactions between water and milk proteins[24]. If the water distribution is homogeneous or if the size of the different water compartments is sufficiently small to allow complete averaging of the two water pools by diffusive exchange then the water relaxation is described by a single mono-exponential curve. On the other hand, if the fast diffusion limit is not verified then the relaxation curve is multi-exponential. The change between mono-exponential and multi-exponential behaviour is not always easy to detect. For example, in a highly concentrated casein gel obtained from rehydration of retentate powder, water relaxation can be described by a narrow relaxation time distribution (Figure 7). This distribution can be explained by slight heterogeneity of the gel induced by imperfect rehydration and mixing. If the gel is warmed at 85°C for 30 minutes, the distribution of the relaxation time increases and this is related to the effect of heat treatment on the water holding capacity of the gel. The amount of water expelled in such a case is not sufficient to induce bi-exponential relaxation, with one relaxing component attributed to the water inside the gel and the other to the water expelled. The relaxation time distribution is highly sensitive to the formation of small water compartments inside the gel and this situation is encountered in young cheeses. After draining and salting the water distribution is non-uniform. Moreover because of the cutting of the curd before draining, small amounts of residual water pockets are entrapped between the curd pieces.

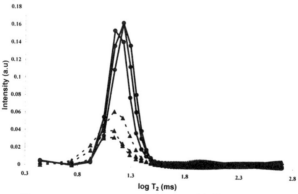

Figure 7 *Effects of heat treatment on T_2 distribution of reconstituted low fat retentate. The T_2 distribution of the untreated sample (●) was enlarged by the heat treatment (▲). The measurements were performed with a 20 MHz spectrometer (Bruker) at 6°C.*

These water compartments lead to increased T_2 distribution if studied by NMR but can also be detected by MRI (Figure 8). Using MRI, the water pocket characterised by longer T_2 values appears in white in the image presented. After the threshold of the image , the water pocket can be quantified according to ripening time (Figure 9).

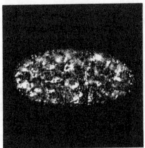

Figure 8 *T_2 weighted MR images of a young soft cheese. The image was acquired with an Open scanner (Siemens) at 0.2 Tesla, with temperature fixed at 5°C.*

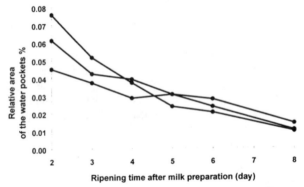

Figure 9 *Relative area of water pockets in a young cheese according to ripening time. The area was calculated after the threshold of the T_2 weighted MR images. The image was acquired with an Open scanner (Siemens) at 0.2 Tesla, with temperature fixed at 5°C.*

Gels formed from milk by renneting may subsequently show syneresis, i.e expel liquid, because the gel contracts[3].Control of the water content during syneresis is complicated because manipulation of the gel modifies the kinetics and the mechanism involved in syneresis. NMR is a particularly useful technique for such a heterogeneous system as it is both non-destructive and non-invasive[25]. During syneresis, water relaxation is described by a bi-modal distribution of the transverse relaxation time. The two water peaks are related to the location of water in the sample. The long T_2 relaxation time reflects the water expelled from the curd and its value is constant over the period of syneresis [10]. The short relaxation time is attributed to the water protons and macromolecule exchangeable protons entrapped in the gel. Three kinds of information are provided by the NMR method: curd volume-decreased kinetics, curd moisture-decreased kinetics and the curd relaxation rate. An example of the kinetics of the curd shrinkage is given in Figure 10. Three replicate syneresis experiments were performed which showed the good level of reproducibility of the method.

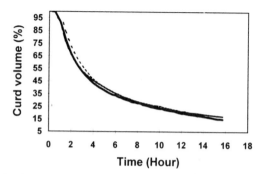

Figure 10 *Curd volume according to time during syneresis. Three replicate kinetics were compared. Measurements were performed with a 10 MHz spectrometer (Bruker).*

The method can be used to study the effects of chemical factors[26] or physical factors such as temperature. For example, curd shrinkage kinetics appear to be sensitive to the ionic strength of milk and to thermal treatment. Indeed, these two factors are known to be important factors in the protein network through the protein-protein interaction, and thus explains their effects on the kinetics and the residual amount of water at the end of syneresis. The casein level in milk only affects the shrinkage rate by an increase in probability of protein-protein bond: the greater the increase in the casein level, the faster the shrinkage.

Parameters of the model	Casein effect	ionic strength effect	Heat treatment effect
y= **a**+bexp(-k*t)	ns*	+	++
y= a+**b**exp(-k*t)	ns	+++	+++
y= a+bexp(-**k***t)	++	++	ns

Table 1 *Statistical analysis of the effects of casein concentration, heat treatment and ionic strength and curd shrinkage kinetics. Casein concentrations were 27 and 37 g/Kg. The ionic strength was reduced to 0.5 compared to that of unmodified milk. Untreated milk and milk heat treated at 72°C for 20 s were compared. The curd shrinkage kinetics were modelled with an exponential equation. Parameter a corresponds to the curd volume at the end of syneresis, b corresponds to volume reduction, and k the rate of shrinkage.*
* for non-significant.

Syneresis and draining can also studied by MRI[27], the advantage being to provide a complete distribution of the water content inside the gel. The method is based on the linear relationship between the water relaxation rate and the water content. A T_2 map is then calculated and converted to the water content map (Figure 11). In this experiment the industrial draining device was directly introduced into the MRI acquisition probehead. This device is a cylinder with a filter fixed at the bottom. The expelled whey can then flow through the curd and the filter to be collected in a container. The container is placed under the draining device.

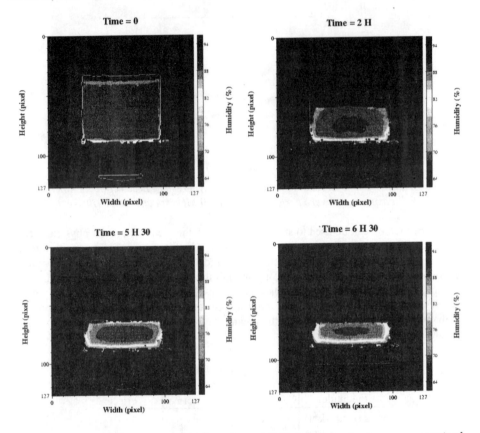

Figure 11 *Curd water distribution according to draining time. Images were acquired with an Open scanner (Siemens) at 0.2 T. A 24 multi-echo sequence was used with TE=17 ms. Temperature was controlled at 5°C.*

In this experiment, a small amount of expelled whey was detected above the curd and in the tank at the beginning of draining. The curd moisture distribution was homogeneous (Figure 11). Two hours later the amount of whey in the tank was greater and we also observed a significant amount of expelled whey above the curd. The curd shrank and the moisture distribution was non-homogeneous. The peripheral area of the curd was dryer than the curd centre and curd moisture heterogeneity increased over time. At the end of draining the moisture range variation between the centre and the peripheral area was 10%. Other information can be extracted from MR images such as the size of the curd, the location of whey, and the amount of whey in the container.

The same approach has been used to study the effects of freeze-thawing on the macroscopic structure and water distribution in the curd[26]. Other MRI techniques can be used for quantitative mapping of moisture or fat in dairy products. For example, a chemical shift selective sequence (CHESS) has been proposed. This sequence provides the fat and water enhanced images separately in cheese[28]. Other MR sequences have been also evaluated on dairy cream[19].

As previously demonstrated, the NMR relaxation time of water is very sensitive to study the dehydration processes of dairy products. The same approach has been used in the case of rehydration of milk powder[29]. The NMR method proposed allows determination of the rehydration kinetics from transverse relaxation time measurements. The method can be used at any temperature and with or without stirring. The variations in relaxation time over time are directly related to the rehydration kinetics. When rehydration is complete, the relaxation time can be related to the casein protein concentration or to the casein structure. This NMR application has been used to differentiate drying process conditions according to the effect on the rehydration capabilities of the powder[30].

4.2 Water dynamics

Information on water dynamics can be obtained from relaxation measurements or with the pulsed field gradient NMR technique. Indeed, relaxation time parameters are sensitive to motion but determination of motion information is model dependent. In contrast to the relaxation method, no assumptions need to be made regarding the relaxation mechanism (s) or in relating the correlation time to the transitional motion of the probe[31]. The first study using the pulsed field gradient NMR technique to measure water diffusion on dairy products was published by Callaghan et al[32].The water self-diffusion coefficients measured in Swiss and Cheddar cheeses were compared at 32°C. No significant difference in the values of the diffusion coefficients was found between the two cheeses. The water self-diffusion coefficient was about one-sixth of the value of bulk water at 30°C and it was suggested that water diffusion is confined to the protein surface. The influence of casein structure and composition on water self-diffusion was recently studied[33]. The results demonstrated surprisingly that the water self-diffusion was not affected by variations between sub-micellar and micellar casein structures. Moreover, water self-diffusion was insensitive to gel formation with rennet. The variations in the water self-diffusion coefficient according to casein concentration was explained by the transport model in porous media. Water diffusion was then explained by two fluxes, one around the micelle and one through the micelle. Inside the micelle the water mobility was probably reduced regionally by low water content and no specific water-protein "binding" was needed to explain the lowering of the water mobility. Other studies are currently in progress on more complex dairy products.

5 FAT

The most widespread NMR application for dairy fat is the solid-fat ratio method. Since this method became an international standard[34], certain improvements have been made. Some have involved measurement of the solid fat ratio in aqueous products such as emulsion[35] and cheese[36] and others have focused on the development of more sophisticated curve-fitting methods for the relaxation curve[37]. In addition, MRI studies have been proposed to quantify the separation of cream from milk[19]

Most of the NMR work on dairy fat has been based on measurement of NMR signal intensity, and little attention has been paid to relaxation time behaviour, especially the relaxation time of the solid fat fraction. The relaxation decay curve of the liquid phase is characterised by a wide distribution of relaxation times[36]. This distribution reflects the complexity of the triglyceride composition and is not very informative. However, the longitudinal relaxation decay of the solid phase is a single exponential curve and potentially more informative regarding the crystal structure of the fat[38]. NMR measurements performed on concentrated emulsions have shown wide variations in relaxation time T_1 between the anhydrous and emulsion state (Table 2). It was shown that when the fat was emulsified the relaxation time of the solid fraction decreased significantly. This effect was dependent on triglyceride fat composition and on the crystal structure. These results have also been reported on real products such as ice-cream[39], and suggest that the characteristics of the crystal structure are modified in the emulsified state. Further studies should be performed to determine the structural scale which should be considered to interpret the relaxation time variations. Such studies should contribute to understanding the effects of emulsion structure and composition of the fat crystal.

	T_1 of the solid fat fraction (ms)	
	bulk state	emulsion state
Vegetable oil	735 (16)	276 (3)
Milk fat	231 (4)	178 (5)

Table 2 *T_1 relaxation time of the solid fat fraction for vegetable oil (cocoa butter and miglyol blend) and milk fat. Measurements were performed at 20 MHz at 4°C.*

The pulsed field gradient NMR method has also been used for determination of droplet size in dairy emulsions[40]. Despite the advantage of the technique for particle sizing, there was been only one report on fat droplet size measurements in cheese[32]. However the technique will be further developed in the factory since the method is now available with commercial bench top NMR spectrometers from Bruker and Resonance Instruments. Since certain NMR experiments are sensitive to motion, flow can be measured with appropriate experimental protocols. These techniques are referred to as NMR velocimetry. The advantage of this technique is that a velocimetry profile or map can be obtained without assumptions concerning the uniformity of the shear rate. For example, the effects of temperature on the rheology of butter fat and margarine has been studied[41]. The results show the effect of temperature on the heterogeneity of the rheological status of the sample.

6 CONCLUSION

As reported by many researchers, numerous length scales, from molecular to macroscopic, can be investigated with NMR and MRI techniques. Until now, different physical techniques have been used to obtain information on this wide range. Because of the different methods used, difficulties have been encountered when links between the different length scales were required. At the same time, dairy technology is becoming more and more sophisticated because of the level of quality required by consumers. Consequently, to answer the question, what length scale should it be considered to explain or control the technological properties are a challenge. NMR manufactures have recently proposed new benchtop equipment with improved facilities for relaxometry, diffusometry and imaging. All these factors will lead to expansion of NMR and MRI applications in dairy science and food manufacturers.

Acknowledgement

The author would like to thank J.P. Innocent and J.M. Soulié from Bongrain Company for very helpful discussions about the fascinating area of cheese technology. I would also like to thank the PhD students who shared with me the exciting work on NMR and MRI in food science. Some of the studies presented in this document were financially supported by the European Community, the French Ministry of Food and Agriculture, the French Association of Dairy Companies "Bretagne Biotechnologies Alimentaires and ARILAIT recherche.

References

1 R.E. Hester and D.E.C. Quine, *J. Dairy Res.*, 1977, **44**, 125.
2 E. Brosio, G. Altobelli and A. Di Nola, *J. Fd. Technol.*, 1984, **19**, 103.
3 P. Walstra, In *Cheese: chemistry, physics and microbiology*, ed. P.F. Fox, Chapman & Hall, London, 1993; p. 141.
4 T.F. Kumosinski, H. Pessen, H.M. Farrell, Jr. and H. Brumberger, *Arch. Bioch.*, 1988, **266**, 548.
5 F. Mariette, C. Tellier, G. Brule and P. Marchal, *J. Dairy Res.*, 1993, **60**, 175.
6 V.D. Fedotov, F.G. Miftakhutdinova and F. Murtazin, *Biofizika*, 1969, **14**, 873.
7 B.P. Hills and S.F. Takacs, *Food Chem.*, 1990, **37**, 95.
8 A.C.M. Van Hooydonk, H.G. Hagedoorn and I.J. Boerrigter, *Neth. Milk Dairy J.*, 1986, **40**, 281.
9 S.P.F.M. Roefs, H. Van As and T. Van Vliet, *J. Food Sci.*, 1989, **54**, 704.
10 F. Mariette and P. Marchal, *J. Magn. Reson Anal.*, 1996, **2**, 290.
11 F. Mariette, P. Maignan and P. Marchal, *Analusis*, 1997, **25**, 24.
12 M.H. Famelart, F. Gaucheron, F. Mariette, Y. Le Graet, K. Raulot and E. Boyaval, *Int. Dairy J.*, 1997, **7**, 325.
13 Y. LeGraet and F. Gaucheron, *J. Dairy Res.*, 1999, **66**, 215.
14 B. Chaland, "Apport de l'IRM bas champ pour l'évaluation des mécansimes d'affinage des fromages pâtes molles et croûtes fleuries". Ph. D. Thesis, University of Rennes, France, 1999.
15 P.F. Fox, J. Law, P.L.H. McSweeney and J. Wallace, In *Cheese: Chemistry, physics and microbiology*, ed. P.F. Fox, Chapman & Hall, London, 1993; p. 389.
16 M. Rosenberg, M.J. McCarthy and R. Kauten, *Food Structure*, 1991, **10**, 185.
17 M. Rosenberg, M.J. McCarthy and R. Kauten, *J. Dairy Sci.*, 1992, **75**, 2083.
18 S.M. Kim, M.J. Mc Carthy and P. Chen, *J. Magn. Reson Anal.*, 1996, **2**, 281.
19 S.L. Duce, M.H.G. Amin, M.A. Horsfield, M. Tyszka and L.D. Hall, *Int. Dairy J.*, 1995, **5**, 311.
20 F. Mariette, G. Collewet, P. Marchal and J.M. Franconi, In *Advances in magnetic resonance in food science*; ed. P. Belton, B. Hills and G. A. Webb, The Royal Society of Chemistry: Cambridge, 1999.
21 F. Mariette, G. Collewet, P. Fortier and J.M. Soulie, In *Magnetic Resonance in food science. a view to the future*, ed. G. A. Webb; P. Belton; A.M. Gil and I. Delgadillo, The Royal Society of Chemistry: Cambridge, 2001.

22 P. Schuck, A. Davenel, F. Mariette, V. Briard, S. Mejean and M. Piot, *Int. Dairy J.*, 2002, **12**, 51.

23 D.G. Dalgleish, *Food Res. Int.*, 1996, **29**, 541.

24 W. Kneifel, A. Seiler, J.N. Dewit and P.S. Kindstedt, *Food Structure*, 1993, **12**, 297.

25 C. Tellier, F. Mariette, J.P. Guillement and P. Marchal, *J. Agric. Food Chem.*, 1993, **41**, 2259.

26 A. Le Dean, "Caracterisation de l'eau dans les produits laitiers: cas des dispersions et des gels egoutes et congeles". Ph. D. Thesis, Institut National Agronomique, Paris, France, 2000.

27 F. Mariette, M. Cambert, F. Franconi and P. Marchal *Les produits alimentaires et l'eau, Agoral 99, Nantes*, 1999, p 405.

28 R. Ruan, K. Chang, P.L. Chen, R.G. Fulcher and E.D. Bastian, *J. Dairy Sci.*, 1998, **80**, 9.

29 A. Davenel, P. Schuck and P. Marchal, *Milchwissenschaft*, 1997, **52**, 35.

30 P. Schuck, "Apprehension des mecanismes de transfert d'eau lors du sechage par atomisation de bases proteiques laitieres et lors de leur rehydratation. Effet de l'environnement glucidique et mineral". Ph. D. Thesis, Ecole National Supérieure d'Agronomie, Rennes, France, 1999.

31 W.S. Price, *Concept Magnetic Resonance*, 1997, **9**, 299.

32 P.T. Callaghan, K.W. Jolley and R. Humphrey, *J. Colloid Interface Sci.*, 1983, **93**, 521.

33 F. Mariette, D. Topgaard, B. Jönsson and O. Söderman, *J. Agric. Food Chem.*, 2002, **50**, 4295.

34 1991.

35 M.A.J.S. Van Boekel, *J. Am. Oil Chem. Soc.*, 1974, **51**, 316.

36 B. Chaland, F. Mariette, P. Marchal and J. de Certaines, *J. Dairy Res.*, 2000, **67**, 609.

37 J. Van Duynhoven, I. Dubourg, G.J. Goudappel and E. Roijers, *J. Am. Oil Chem. Soc.*, 2002, **79**, 383.

38 I.J. Colquhoun and A. Grant *Eurofood Chem V*, Versailles, France, 1989; Vol. 2, p 668.

39 T. Lucas, S. Dominiawsyk, F. Mariette and G. Alvarez *Engineering Food, Eighth ICEF International Conference Proceedings*, Puebla, Mexico; CRC Press; 2002; Vol. 1, p 551.

40 B. Balinov, O. Soderman and T. Warnheim, *J. Am. Oil Chem. Soc.*, 1994, **71**, 513.

41 M.M. Britton and P.T. Callaghan, *J. Texture Stud.*, 2000, **31**, 245.

AN EFFORT TO DEVELOP AN ANALYTICAL METHOD TO DETECT ADULTERATION OF OLIVE OIL BY HAZELNUT OIL

I. Kyrikou[1], M. Zervou[1], P. Petrakis[2], T. Mavromoustakos[1]

[1]National Hellenic Research Foundation, Institute of Organic and Pharmaceutical Chemistry, Vas. Constantinou 48, 11635 Athens, Greece
[2]National Agricultural Research Foundation, Mediterranean Forest Research Institute, Laboratory of Entomology, Alkmanos, 11528 Ilissia, Athens Greece

1 INTRODUCTION

The considerable rise in consumer interest in olive oil as a healthful product and the economic premium attached, particularly to the highest-quality categories, increase the risks of adulteration with cheaper products.

Several analytical methods have been developed to detect adulteration of virgin olive oil by other oils which mainly present different chemical profile. These can be classified as destructive and non-destructive methods. In the destructive methods, the sample for analysis is subjected to a chemical treatment while in a non-destructive method the sample is free of any treatment. [1-6]

Chromatographic methods currently used to detect adulteration of virgin olive oil (VO) by other oils suffer from the disadvantage that they are destructive methods. They are also time-consuming, not specific and qualitative ones. Therefore, a great effort is recently made to develop analytical methodologies that do not suffer from the above dissadvantages. These methodologies are based either on the different chemical composition of the major or minor constituents of olive oil relatively to other oils.

We have previously applied ^{13}C nuclear magnetic resonance (NMR) to analyze quantitatively the most abundant fatty acids in olive oil. We have reported that this analysis differed only by up to 5% compared to the corresponding gas chromatography analysis. [7] Recently, we applied quantitative analysis on the olefinic region of the high resolution ^{13}C NMR spectrum of virgin olive oil in order to detect its adulteration by other oils that differ significantly in the chemical profile of fatty acids. [8]

This manuscript reports preliminary results regarding the application of the developed methodology in samples containing mixtures of olive oil of different origin, quality and variety with hazelnut oil. The adulteration of olive oil by hazelnut oil is more challenging since both oils have similar chemical profile.

2. METHOD AND RESULTS

2.1   Sample preparation

The sample containing about 1 mL of olive oil or a mixture of olive oil with hazelnut oil in $CDCl_3$ (40% w/w) with a known mass of 1,4 pyrazine as an internal reference were added in a 5-mm NMR tube and mixed thoroughly. Spectra were recorded on a Bruker AC 300 spectrometer at a temperature of 313 K. The inverse gate pulse program was used to give fully decoupled spectra with no NOE. The olefinic (128-138 ppm) and carbonylic regions (-170 ppm) were subjected to quantitative analysis by comparing the area of the observed peaks with the area of pyrazine peak which corresponds to a known mass content expressed in mmol. Pyrazine gives only a single peak at a chemical shift of 144.8 ppm and does not coincide with any of the observed peaks. Peaks are simulated using Lorentzian lineshapes provided by Bruker WIN-NMR s/w.

The mass content of a given peak is normalized to 1g of oil. The formula used to express the results in mmol/g is

$$\chi_i(mmol/g) = 4\,(Wp/Mrp)\,(Si/Sp)\,1000/Woil \tag{1}$$

where Wp(mg) is the weight of pyrazine added, Mrp is the molecular weight of pyrazine, Woil (mg) is the weight of oil (or mixture of oils), Si is the area of the individual peak calculated after deconvolution of the peaks and Sp is the area of the pyrazine calculated also after integration. The factor four is used to normalize over the four magnetically equivalent pyrazine carbon atoms.

2.2   Experimental Results

Representative high resolution ^{13}C NMR spectra of samples containing pure olive and hazelnut oils of different origin and quality are shown in Figures 1 and 2 (olefinic and carbonyl regions respectively). The olefinic region consists of 12 peaks corresponding to the oleic (O) and linoleic (L) moieties present in α and β positions of the glycerol backbone, namely O-9α, O-9β, O-10α, O-10β, L-9α, L-9β, L-10α, L-10β, L-12α, L-12β, L-13α, L-13β. From the observed peaks in the carbonylic region only the peak corresponding to saturate acids labeled as Sat is subjected to quantitative analysis. The rest of the peaks due to oleic and linoleic acids are rejected due to serious overlapping.

Figure 3 shows the olefinic region of mixtures containing olive oil and hazelnut oil and Figure 4 shows the carbonyl region of the same mixtures respectively. The chemical shifts changes of these peaks are given in Figure 5.

Qualitative analysis of these data shows that the chemical shifts and integration of the peaks can be diagnostic parameters only if statistical analysis is sought in a more expanded database. We are currently performing more experiments including various mixtures (up to 50% adulteration level) in order to define the detection limit of adulteration and define the factors responsible for the detection limit. Such factors may be the geographic origin (i.e. samples from different countries Italy, Greece, Spain), the variety of the grade of the samples analysed (i.e. refined or lampante) or variety of oils (olive oil, hazelnut oil). Thus, taxonomy, geographic origin and oil quality in the samples analyzed may create a nest effect that limits the adulteration level. A preliminary results using GMDSCAL unsupervised statistical analysis showed that the created dendrogram consists of various defined regions. Although these results are promising indicate that indeed the above-mentioned factors may shadow the easy detection of adulteration. The degree of interference of these factors remains to be established with further experiments.

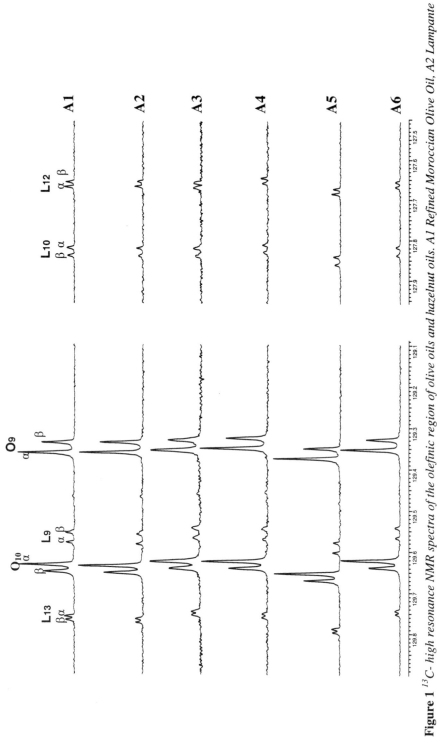

Figure 1 ^{13}C- high resonance NMR spectra of the olefinic region of olive oils and hazelnut oils. A1 Refined Moroccian Olive Oil, A2 Lampante Greek Olive Oil, A3 Refined Italian Pomace Oil, A4 Refined Spanish Olive Oil, A5 Refined French Hazelnut Oil, A6 Virgin Spanish Hazelnut Oil

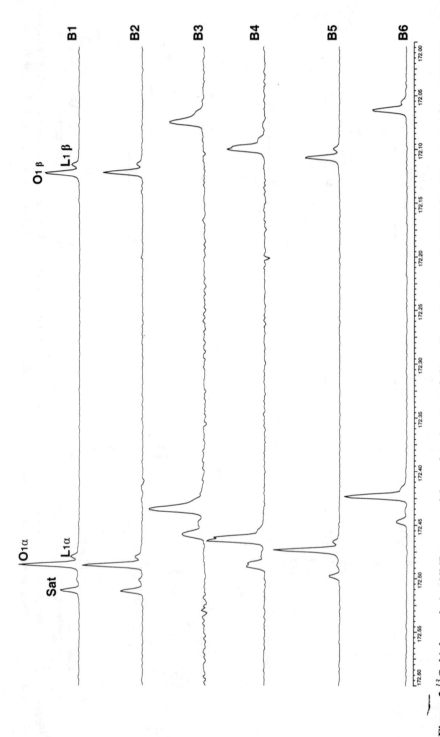

Figure 2 ^{13}C- high resolution NMR spectra of the carbonyl region of olive oils and hazelnut oils. B1 Refined Moroccian Olive Oil, B2 Lampante Greek Olive Oil, B3 Refined Italian Pomace Oil, B4 Refined Spanish Olive Oil, B5 Refined French Hazelnut Oil, B6 Virgin Spanish Hazelnut Oil

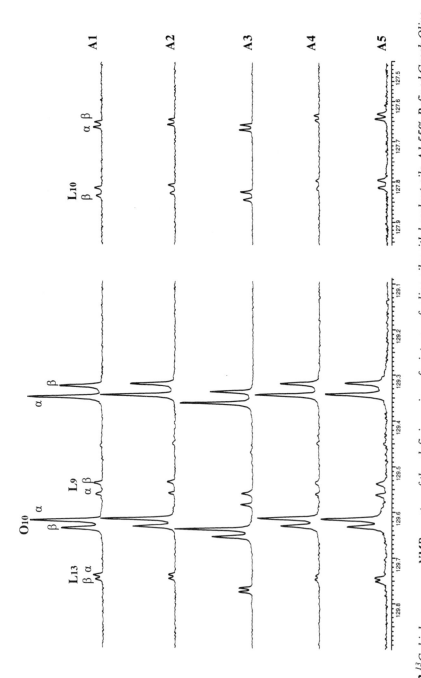

Figure 3 ^{13}C- high resonance NMR spectra of the olefinic region of mixtures of olive oils with hazelnut oils. A1 55% Refined Greek Olive Oil + 45% Refined Turkish Hazelnut Oil, A2 70% Refined Greek Olive Oil + 30% Refined French Hazelnut Oil, A3 80% Refined Moroccian Olive Oil + 20% Refined Turkish Hazelnut Oil, A4 86% Lampante Spanish Olive Oil + 14% Virgin Spanish Hazelnut Oil, A5 89% Refined Italian Olive Oil + 11% Refined Turkish Hazelnut Oil

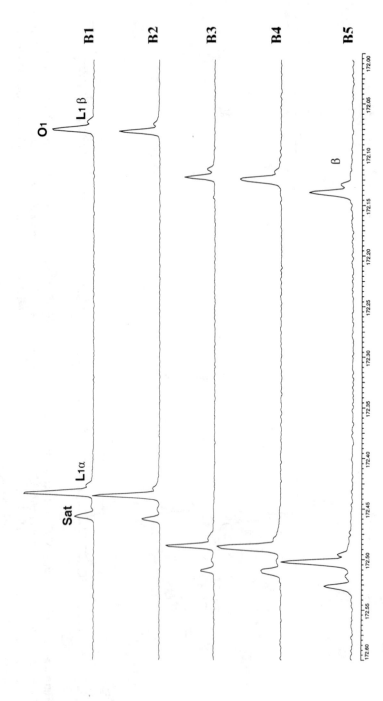

Figure 4 *¹³C- high resolution NMR spectra of the carbonyl region of mixtures of olive oils with hazelnut oils. B1 55% Refined Greek Olive Oil + 45% Refined Turkish Hazelnut Oil, B2 70% Refined Greek Olive Oil + 30% Refined French Hazelnut Oil, B3 80% Refined Moroccian Olive Oil + 20% Refined Turkish Hazelnut Oil, B4 86% Lampante Spanish Olive Oil + 14% Virgin Spanish Hazelnut Oil, B5 89% Refined Italian Olive Oil + 11% Refined Turkish Hazelnut Oil*

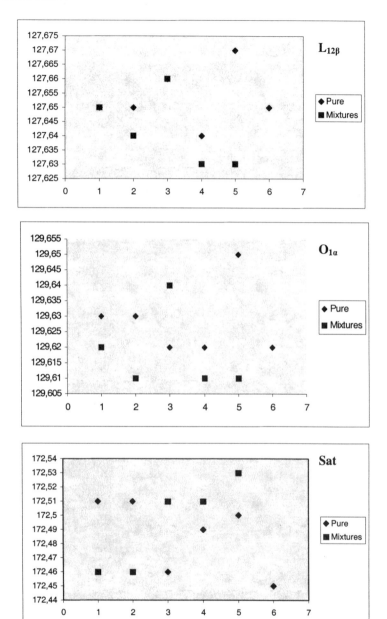

Figure 5 *Chemical shift changes of representative peaks for the pure oils and the mixtures.*
Pure: 1. Refined Moroccian Olive Oil, 2. Lampante Greek Olive Oil, 3. Refined Italian Pomace Oil, 4. Refined Spanish Olive Oil, 5. Refined French Hazelnut Oil, 6. Virgin Spanish Hazelnut Oil. Mixtures: 1. 55% Refined Greek Olive Oil + 45% Refined Turkish Hazelnut Oil, 2. 70% Refined Greek Olive Oil + 30% Refined French Hazelnut Oil, 3. 80% Refined Moroccian Olive Oil + 20% Refined Turkish Hazelnut Oil, 4. 86% Lampante Spanish Olive Oil + 14% Virgin Spanish Hazelnut Oil, 5. 89% Refined Italian Olive Oil + 11% Refined Turkish Hazelnut Oil

References

1. M.L. Ruiz del Castilo, M. Herraiz, M.D. Molero, A. Herrera. *JAOCS*, 2001,**78**,1261.
2. L. Mannina, M. Patumi, P. Fiordiponti, M.C. Emanuele, A.L. Segre. *Ital. J. Food Sci.* 1999, **11**,139.
3. J. Spangenberg, S. Macko, J. Hunziker. *J. Agric. Food Chem.* 1998, **46**, 4179.
4. C. Fauhl, F. Reniero, C. Guillou. *Magn. Reson. Chem.* 2000, **38**, 436.
5. R. Apariccio, M.T. Morales. *J. Agric. Food Chem.* 1998, **46**, 1116.
6. G. Vlahov. *JAOCS,* 1999, **76**, 1223.
7. T. Mavromoustakos, M. Zervou, E. Theodoropoulou, D. Panagiotopoulos, G. Bonas, A. Helmis. *Magn. Reson. Chem.* 1995, **35**:S3.
8. T. Mavromoustakos, M. Zervou, G. Bonas, A. Kolocouris, P. Petrakis. *JAOCS*, 2000, **77**,405.

QUALITY CONTROL OF VEGETABLE OILS BY ^{13}C NMR SPECTROSCOPY

Rosario Zamora, Gemma Gómez, and Francisco J. Hidalgo

Instituto de la Grasa, CSIC, Avenida Padre García Tejero 4, 41012 Sevilla, Spain

1 INTRODUCTION

Nowadays, the analysis of vegetable oils is dominated by classic determinations, such as acidity, peroxide value, ultraviolet absorption, etcetera, as well as by the use of chromatographic procedures, including thin layer, gas, and high performance liquid chromatography.[1,2] These techniques are primarily used for quantitative measurement of known compounds, and with these analytical criteria different international regulations have been developed to define oil genuineness and quality.

One drawback to these procedures is that there are too many different assays to be applied to routine quality control. In addition, many of these methods require the isolation and analysis of minor compounds from the unsaponifiable matter by means of procedures that are laborious and time-consuming. Therefore, many studies have been carried out to apply new analytical techniques that, with very little or no manipulation of the sample, can give results similar or superior to those obtained by the classical procedures.

In this context, instrumental spectroscopic techniques, and particularly high resolution NMR, have emerged as potential tools in recent years. Thus, NMR has been applied to the determination of genuineness, quality, and geographical and varietal origin identification.[3,4] However, NMR has also limitations, which are mostly related to the oil composition. Oils are composed of more than 95 % of triacylglycerols, and obtaining information about components other than triacylglycerols usually requires extended acquisition times and minimal overlap of their NMR signals with those of triacylglycerols.

In an attempt to overcome these limitations, some years ago, Zamora et al.[5] studied the use of the unsaponifiables of the oils for defining oil authenticity and genuineness with promising results. However, the use of unsaponifiables has also several limitations: it is a time-consuming process, and the structural information on triacylglycerols and other saponifiable components is lost.

As an alternative procedure, Zamora et al.[6] have recently concentrated the minor components of the oils by means of an easy chromatographic procedure that was used satisfactorily in oil characterization.[7] Because the fractions obtained contain most of the oil components that contribute to oil quality in a higher concentration than original oils, it should be expected that ^{13}C NMR analysis of these fractions could be employed to determine oil quality much more satisfactorily than employing complete oils. As representative measurements of oil quality, the determination of stability (which is related to the shelf life of the oils), polar compounds (related to thermal, oxidative, and hydrolytic alterations of the oils), and oil color (which plays a significant role in consumers' acceptability) were selected.

2 MATERIALS AND METHODS

2.1 Materials

The oils employed in this study included virgin olive oils from different cultivars and regions of Europe and north Africa, "lampante" olive oils, and different refined oils, including olive, olive pomace, hazelnut, low-erucic rapeseed, high-oleic sunflower, corn, grapeseed, soybean, and sunflower oils. Most of the samples were obtained from our Institute's experimental oil mill (Instituto de la Grasa, Sevilla, Spain), the Institute's Department of Analysis, the Institute's Pilot Plant, or Koipe S. A. (Andujar, Jaén, Spain). In addition, some of the refined oils were prepared and refined in our laboratory using a laboratory-scale apparatus described previously by Dobarganes et al.[8] This procedure included degumming with phosphoric acid, neutralization with sodium hydroxide, bleaching with bleaching earth (Trisyl) for 10 min at 90 °C, and deodorization under vacuum (1 mm) at 250 °C for 3 h.

2.2 Oil Fractionation

Triplicate samples of the oils were fractionated by column chromatography using 19 g of silica gel 60 (particle size 0.063-0.200 mm) as absorbent, which was obtained from Merck (Darmstadt, Germany) and used without any previous treatment. The column was prepared in the elution solvent: a mixture of hexane and diethyl ether (87:13). The oil (6 g) was dissolved in 10 mL of the same solvent and introduced in the column. Most nonpolar compounds, including a significant portion of the triacylglycerols, were eluted with 100 mL of the elution solvent and were discarded. The oil fraction, containing polar compounds as well as a small part of triacylglycerols and other nonpolar compounds, was then eluted with 100 mL of acetone.

2.3 Oil Analysis

Oils were analyzed to determine their stability, color, and polar compounds content and composition. Oil stability was determined by the Rancimat method (Metrohm Co., Basel, Switzerland) at 110 °C. Polar compounds were determined by solid-phase extraction and size-exclusion chromatography according to Márquez-Ruiz et al.[9] Color was determined directly using filtered oils on a Shimadzu UV-2401 PC spectrophotometer by means of a software for color analysis.

2.4 NMR Spectroscopy

^{13}C NMR spectroscopy was performed on a Bruker AC 300P (Bruker Instruments, Inc., Karlsruhe, Germany) operating at 75.4 MHz. The oil fractions obtained by column chromatography as described above were evaporated, dissolved in 700 μL of $CDCl_3$ (containing 0.03 % vol/vol tetramethylsilane), introduced into a 5-mm NMR tube, and acquired as described previously.[7] Free induction decays were transformed by using absolute intensity, and chemical shifts were related to the signal for tetramethylsilane (δ 0

ppm). The solvent CDCl₃ was used as the internal standard for height intensity and to correct for small changes in field homogeneity. One hundred and thirty-five peaks at the same chemical shifts/positions were selected, and peaks heights were recorded for use in the data analysis of intensity patterns. The recorded intensities for each oil were collected in a matrix, with each row containing all 135 peaks of one spectrum. This matrix also included oil stability, polar compounds, and color values, which were determined as described above. The values used in the data analysis were the average of the three replicates obtained for each oil. No further preprocessing of the data was performed.

2.5 Data Analysis

Statistical data analysis was performed with the SPSS for Windows (version 11.0.1) statistical package (SPSS Inc., Chicago, IL). Stepwise linear regression analysis (SLRA) was applied to the data matix, prepared as described above, to select the variables (NMR signals) that better explained the oil stability determined by the Rancimat, the polar compounds determined by HPSEC, and the oil color determined spectrophotometrically.

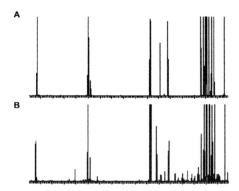

Figure 1 *¹³C NMR spectra of: A, a virgin olive oil, and B, the fraction enriched in polar compounds of the same oil. Samples were dissolved in CDCl₃ and acquired for: A, 1 h, or B, 4 h.*

3 RESULTS

3.1 ¹³C NMR Spectra of Oil Fractions

¹³C NMR spectra of oil fractions were much more complex than those obtained for complete oils. Figure 1 shows the NMR spectra of a virgin olive oil and the corresponding spectra of the fraction enriched in polar compounds. The spectra of oil fractions exhibited carbonyl carbons between 177.8 and 172.8 ppm (signals P1-P13), olefinic carbons between 141.0 and 121.0 ppm (signals P14-P29), glycerol carbons between 72.1 and 61.0 ppm (signals P34-P44) and aliphatic carbons between 58.0 and 11.8 ppm (signals P45-P135). These signals corresponded to the carbon atoms of the different components present in the isolated fractions: triacylglycerols, polymeric triacylglycerols, oxidized triacylglycerols, diacylglycerols, monoacylglycerols, and free fatty acids, as well as the various minor polar components of the oils: sterols, fatty alcohols, phenols, etc. A detailed analysis of obtained

spectra and the assignations of the corresponding signals was described previously.[7] A selected list of signals of interest in the present study along with their corresponding assignations are collected in Table 1.

Table 1   *Chemical Shifts and Assignations of 13C NMR Signals Selected by SLRA*

Signal	δ (ppm)	Assignment[a]	Signal	δ (ppm)	Assignment[a]
P1	177.75	CA (FFA1)	P69	37.24	ST1
P2	174.32	CA (1MAG1α)	P72	36.13	STchain
P3	173.94	CA (1,3DAG1α)	P75	35.31	UK
P10	173.31	CA (TAG1α)	P78	34.06	S2α
P12	172.90	CA (TAG1β)	P79	34.00	O2α; L2α; ST
P15	132.73	OL	P80	31.93	Sω3
P17	130.17	OL (L13)	P82	31.86	ST
P18	129.97	OL (O10; L9)	P85	31.48	UK
P21	128.82	OL	P87	29.76	ME
P23	128.18	OL	P88	29.70	ME
P28	125.31	OL	P91	29.60	ME
P30	86.79	UK	P94	29.37	ME
P32	85.50	UK	P98	29.09	ME
P33	78.89	UK	P99	29.03	ST
P34	72.04	GL (1,2DAG2)	P100	28.99	UK
P36	70.17	GL (1MAG2)	P101	28.25	ST
P37	69.96	UK	P102	27.79	UK
P38	68.85	GL (TAG2)	P103	27.33	UK
P39	68.06	GL (1,3DAG2)	P105	27.18	AL (L8; L14)
P40	64.99	GL (1,3DAG1/3; 1MAG3)	P106	27.16	AL (O8)
P41	63.31	GL (1MAG1; FH1)	P112	25.48	UK
P42	62.17	GL (1,2DAG3; 2MAG1/3)	P115	24.28	ST
P52	52.25	UK	P121	21.83	UK
P56	48.03	UK	P122	21.06	ST
P57	47.08	UK	P125	19.81	ST
P61	42.28	ST13	P128	18.99	ST
P62	42.14	ST4	P133	14.08	Lω1
P65	39.32	UK	P134	11.94	ST

[a] Assignments are abbreviated with carbon type followed by compound and carbon number, if known. AL, allylic; CA, carbonyl; DAG, diacylglycerol; FFA, free fatty acid; FH, fatty alcohol; GL, glycerol; L, linoleic; ME, methylene envelope; MAG, monoacylglycerol; O, oleic; OL, olefinic; S, saturated; ST, sterol; TAG, triacylglycerol; UK, unknown.

3.2 Oil Stability Prediction by ^{13}C NMR

By using SLRA, it was possible to select the ^{13}C NMR signals that better explained the oil stability determined by the Rancimat. When applied to the sixty-six oils analyzed (36 virgin olive, 1 "lampante" olive, 3 refined olive, 6 refined olive pomace, 2 low-erucic rapeseed, 3 high-oleic sunflower, 3 corn, 3 grapeseed, 4 soybean, and 5 sunflower oils), stability could be predicted (r = 0.914, *p* < 0.0001) by using intensities of P1, P18, P52, P56, P78, P79, P82, and P103 signals. These signals are related to the type and content of fatty acids, triacylglycerols, recognized antioxidant compounds as well as other unidentified oil components. Stability prediction by using NMR data was better than stability predicted using chemical determinations (fatty acid, triacylglycerol, phenol, and tocopherol contents and compositions).[10] Figure 2 shows the plot of the stabilities calculated for the sixty-six analyzed oil samples against the stabilities determined by the Rancimat.

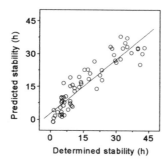

Figure 2 *Plot of stabilities determined by the Rancimat method against stabilities predicted by stepwise linear regression analysis using the* [13]*C NMR data obtained from sixty-six oil fractions.*

3.3 Polar Compound Determination by ^{13}C NMR

Analogously to stability prediction, SLRA also allowed to select the ^{13}C NMR signals that better explained the different polar compounds determined in the oils by high performance size exclusion chromatography (HPSEC). When applied to the eighty-seven oils analyzed (41 virgin olive, 7 "lampante" olive, 5 refined olive, 6 refined olive pomace, 3 hazelnut, 3 low-erucic rapeseed, 4 high-oleic sunflower, 4 corn, 3 grapeseed, 5 soybean, and 6 sunflower oils), polymeric triacylglycerol, oxidized triacylglycerol, diacylglycerol and free fatty acid contents determined by ^{13}C NMR were correlated with those values determined by HPSEC (r = 0.962, *p* < 0.0001; r = 0.957, *p* < 0.0001; r = 0.960, *p* < 0.0001; and r = 0.872, *p* < 0.0001, respectively). SLRA selected 18 signals for polymeric triacylglycerols (P10, P12, P17, P21, P23, P33, P34, P39, P57, P75, P87, P88, P98, P100, P102, P112, P121, and P133), 8 for oxidized triacylglycerols (P18, P30, P37, P38, P78, P80, P85, and P102), 8 for diacylglycerols (P34, P38, P40, P69, P79, P101, P105, and P115), and 7 for free fatty acids (P1, P2, P62, P87, P99, P106, P134). All these signals were related to the different components present in the isolated fractions. Figure 3 shows the plot of the polar compounds calculated by using NMR data against the polar compounds content determined by HPSEC.

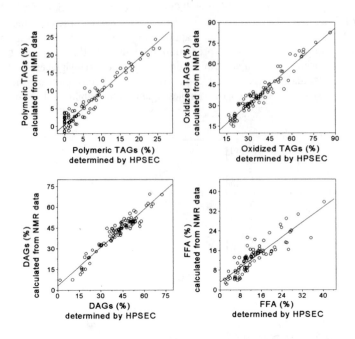

Figure 3 *Plots of the contents of polymeric triacylglycerols (TAGs), oxidized TAGs, diacylglycerols (DAGs), and free fatty acids determined by high performance size exclusion chromatography (HPSEC) against the contents of the same compounds calculated from NMR data.*

3.4 Color Determination by ^{13}C NMR

As described above for stability prediction and polar compounds determination, SLRA also allowed to select the ^{13}C NMR signals that better explained the oil color determined using the CIELAB system.[11] When applied to the eighty-five oils analyzed (41 virgin olive, 7 "lampante" olive, 3 refined olive, 6 refined olive pomace, 2 hazelnut, 3 low-erucic rapeseed, 4 high-oleic sunflower, 4 corn, 4 grapeseed, 5 soybean, and 6 sunflower oils), L^*, a^*, and b^*, as well as yellowness index (YI, expressed according to Francis and Clydesdale[12] as a ratio between b^* and L^*) determined by ^{13}C NMR were correlated with those values determined spectrophotometrically ($r = 0.863$, $p < 0.0001$; $r = 0.783$, $p < 0.0001$; $r = 0.932$, $p < 0.0001$; and $r = 0.967$, $p < 0.0001$, respectively). SLRA selected 10 signals for L^* (P34, P36, P38, P40, P42, P61, P65, P88, P115, and P125), 7 for a^* (P36, P57, P72, P94, P106, P122, and P128), 10 for b^* (P2, P10, P30, P40, P41, P52, P88, P98, P100, and P105), and 19 for YI (P1, P3, P10, P12, P15, P28, P30, P32, P34, P40, P41, P65, P69, P75, P88, P91, P98, P105, and P128). All these signals were related to the different components present in the isolated fractions. Figure 4 shows the plots of L^*, a^*, b^*, and YI values calculated by using NMR data against L^*, a^*, b^*, and YI values determined spectrophotometrically.

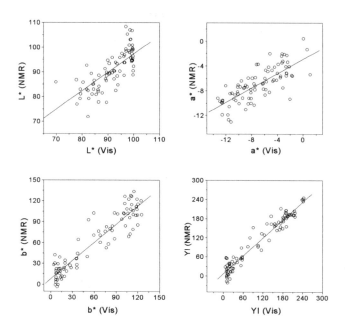

Figure 4 *Plots of L*, a*, b*, and yellowness index (YI) determined spectrophotometrically against L*, a*, b*, and YI values calculated from NMR data.*

4 DISCUSSION

Nowadays, the main limit for a routine application of NMR spectroscopy to oil analysis is the usual low concentration of the compounds which define the genuineness, origin and quality of vegetable oils. Thus, for example, phenols or tocopherols, which play a significant role in oil stability, or chlorophylls and carotenes, which are mostly responsible for oil colors, are usually under the detection limits of common conditions in NMR spectroscopy. Therefore, it should be expected that an increase in the concentration of all these minor compounds in the oils would increase considerably the possibilities of application of NMR to routine oil analysis.

The results obtained in this study confirm this hypothesis and provide a methodology that may be easily applied to obtain in a short time period much of the information needed to define oil genuineness and quality. Thus, the chromatographic fractionation of the oils eliminated a significant part of the triacylglycerols and produced fractions enriched in polar compounds which were adequately employed to determine oil stability, color, and polar compounds content and composition as examples of different measurements usually carried out for determining oil quality. These parameters, which are usually analyzed by means of different procedures, some of which are laborious and time-consuming, may be determined simultaneously by NMR, at the same time that information on oil genuineness is also acquired. Analogously to the better results attained for oil genuineness determination by using oil fractions[7] than by using complete oils,[4] the oil quality results obtained in this study should not be expected when using complete oils.

All these results suggest that the chromatographic fractionation of oils is a necessary preliminary step in NMR application to oil analysis, in spite of the difficulties added with the introduction of this step.

5 CONCLUSIONS

The study by ^{13}C NMR spectroscopy of fractions obtained chromatographically from vegetable oils provides with only one measurement information on very different aspects of oil quality that nowadays need many different analyses. The isolation of these fractions, which may also be employed for oil genuineness determination by ^{13}C NMR, seems to be a necessary preliminary step to achieve a routine application of NMR to oil analysis.

Acknowledgments

We are indebted to Koipe S.A. for the gift of many of the oils used in this study, to Prof. M.C. Dobarganes for the HPSEC data, and to J.L. Navarro y M. Giménez for the technical assistance. This study was supported in part by the European Union, the Instituto Nacional de Investigación y Tecnología Agraria y Alimentaria (Project CAO98-008), and the Comisión Interministerial de Ciencia y Tecnología of Spain (Project 1FD97-0343).

References

1   R.J. Hamilton, ed., *Lipid Analysis in Oils and Fats*, Blackie, London, 1998.
2   C.C. Akoh and D.B. Min, eds., *Food Lipids. Chemistry, Nutrition, and Biotechnology*, 2nd ed., Marcel Dekker, New York, 2002.
3   R. Sacchi, 'High Resolution NMR of Virgin Olive Oil' in *Magnetic Resonance in Food Science. A view to the Future*, eds., G.A. Webb, P.S. Belton, A.M. Gil and I. Delgadillo, Royal Society of Chemistry, Cambridge, 2001, pp. 213–226.
4   R. Zamora, V. Alba and F.J. Hidalgo, *J. Am. Oil Chem. Soc.*, 2001, **78**, 89.
5   R. Zamora, J.L. Navarro and F.J. Hidalgo, *J. Am. Oil Chem. Soc.*, 1994, **71**, 361.
6   R. Zamora, G. Gómez, M.C. Dobarganes and F.J. Hidalgo, *J. Am. Oil Chem. Soc.*, 2002, **79**, 261.
7   R. Zamora, G. Gómez and F.J. Hidalgo, *J. Am. Oil Chem. Soc.*, 2002, **79**, 267.
8   M.C. Dobarganes, M.C. Pérez-Camino, G. Márquez-Ruiz and M.V. Ruiz-Méndez, 'New Analytical Possibilities in Quality Evaluation of Refined Oils' in *Edible Fats and Oil Processing: Basic Principles and Modern Practices*, ed., D.R. Erikson, American Oil Chemists' Society, Champaign, Illinois, 1990, pp. 427-429.
9   G. Márquez-Ruiz, N. Jorge, M. Martín-Polvillo and M.C. Dobarganes. *J. Chromatogr. A*, 1996, **749**, 55.
10 F.J. Hidalgo, G. Gómez, J.L. Navarro and R. Zamora. *J. Agric. Food Chem.*, 2002, **50**, in press.
11 CIE (Commission Internationale de l'Eclairage), *Recommendations on Uniform Color Spaces, Color-Differences Equations, Psychometric Color Terms*, CIE Publication 15 (Supplement 2), Bureau Central de Commission Internationale de l'Eclairage, Paris, France, 1978.
12 F.J. Francis and F.H. Clydesdade, *Food Colorimetry: Theory and Applications*, AVI Publishing, Westport, Connecticut, 1975.

MULTIVARIATE ANALYSIS OF TIME DOMAIN NMR SIGNALS IN RELATION TO FOOD QUALITY

Elisabeth Micklander[1], Lisbeth G. Thygesen[1], Henrik T. Pedersen[1], Frans van den Berg[1], Rasmus Bro[1], Douglas N. Rutledge[2] and Søren B. Engelsen[1]

[1]Center for Advanced Food Studies, Food Technology, Department of Dairy and Food Science, The Royal Veterinary and Agricultural University, Rolighedsvej 30, 1958 Frederiksberg C, Denmark
[2]Institut National Agronomique, Laboratoire de Chimie Analytique, 16, rue Claude Bernard, 75005 Paris, Cedex 05, France

1 INTRODUCTION

NMR is one of the most successful analytical techniques of our time. It is directly applicable to amorphous, heterogeneous and opaque systems such as food and foodstuffs. Time domain NMR (TD-NMR) is a relatively inexpensive version of the technique, which still has the potential to serve as a unique window into the complicated food matrix. It allows food engineers and scientists to "get in touch" with the protons in a sample. The major food industries all have problems for which the best method of measurement is TD-NMR. These include, for example, **solid fat** determination in the plant oil industry, **total fat** content in the slaughter, oil seed, fishmeal, confectionary and dairy industry, **droplet size** in the margarine/butter/spread industry, **adsorbed and total water** in the feed, ingredients, snacks and cookie industries and **gel-formation** in the stabilizer/hydrocolloid industry. In addition to these traditional applications TD-NMR offers the possibility to explore end-properties of foods such as texture, mouth feel and slipperiness to the extent that these features depend on the compartmentalization and mobility of water.

The TD-NMR technique is one the most versatile, multivariate, low perturbation, non-destructive analytical techniques on the market. It gives volume measures that in contrast to optical methods are insensitive to surface characteristics and can handle highly opaque samples. Despite the fact that the TD-NMR signals are multivariate, industrial use tends to be strictly application limited, normative and univariate. The reasons for this apparent misuse of TD-NMR are multi-facetted, but one problem is that TD-NMR is associated with a difficult and "dusty" image (poor compatibility with the user), a general lack of university training and instrument manufacturers that have developed in-house software that constrains the users rather than opening the way to the immense possibilities offered by the technique.

TD-NMR has disadvantages as well. Firstly a strong magnetic field is required. This is a technological problem that can be solved even near process lines, for example by one-sided NMR instruments. Secondly, instrument initialization, pulse settings and data acquisition give infinite choices, which in the first place would appear as an advantage, but in practice is a hindrance. Signals from TD-NMR instruments are of relatively low resolution and a few pre-constructed and pre-optimized ready-to-use pulse sequences will be able to capture all necessary NMR information. Thirdly, because TD-NMR is relatively low resolution, the few functional elements or proton populations in the relaxation decays all have to be found within the same narrow time-constant range. This fact makes data interpretation difficult and tendentious. A good example is the interpretation of TD-NMR signals from meat, which despite relative simplicity and numerous studies still are not fully elucidated[1-3].

The quantitative aspect of TD-NMR data analysis has and still is suffering from over-simplified data analysis, mainly because the complexity is too high for a deterministic approach and because it is conceptually difficult to cope with a high number of variables. This simplified tradition has resulted in the widespread normative use of TD-NMR: if you keep only the part of the data that you think is relevant, you will confirm what you already know is important and thus reduce your chances of innovation.

We believe that NMR scientists in industry and academia can learn from the successful progress of Near-infrared (NIR) spectroscopy, which in the last decade has revolutionized quality control in practically all areas of primary food and feed production. Near-infrared spectroscopy has been implemented for monitoring quality of millions of samples of cereals, milk and meat with unprecedented precision and speed. The key to this success is the extraordinary synergy that lies in the merging of a rapid non-invasive spectroscopic sensor and the new data technology called chemometrics. In this paper we have therefore set out to investigate the explorative potential of TD-NMR via the application of multivariate data analysis (chemometrics). Using this approach engineers in the food industry are now able to explore their processes using so-called soft models and spectroscopic sensors to complement traditional deterministic models and univariate measurements. This dramatic development appears to transform the food industry from a traditionally low-technology industry into a high-technology one, alongside quality control developments in the pharmaceutical industry. This technology leap, while still in its onset, has by now matured to a stage where the exploratory TD-NMR approach should also be included in the toolbox of food researchers and scientists.

2 EXPLORATORY DATA ANAYLYSIS

Perhaps because NMR was developed by physicists and chemists, it has primarily been used for identification, authenticity determination and simple quantification purposes. Its explorative potential has largely remained unexploited. The first report on the use of chemometrics on NMR data appeared in 1983 by Johnels *et al.*[4], but it was not before the early nineties that new developments of applying multivariate data analysis to high-resolution NMR signals occurred, with the impetus being the analysis of complex 1D and 2D NMR structures of complex biological and pharmaceutic/therapeutic matrices[5,6]. The use of chemometric data analysis to TD-NMR which is still in its infancy, was first adapted in the late nineties[7-9].

Visualization (certainly!) **and Data Mining** (maybe?)

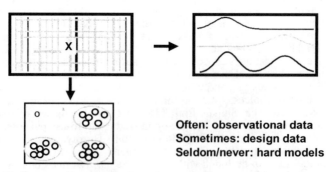

Often: observational data
Sometimes: design data
Seldom/never: hard models

Figure 1   *Exploratory Data Analysis*

One of the main advantages of chemometric data analysis is the possibility of projecting multivariate data into a few dimensions via a graphical representation. Chemometrics is also called "statistics without tears", because it makes it possible to handle large data sets and deals efficiently with real-world multivariate data, taking advantage of the previously feared co-linearity of spectral data. With chemometrics it is even possible to analyse whole spectra in real time. The chemometric methods are based on the calculation of underlying latent data structures using a two-dimensional data strategy, i.e. measuring a series of samples and finding common latent data structures (Figure 1). In this paper we will demonstrate the application of multivariate data analysis to a number of different food systems and through those evaluate the current status of the use of chemometrics for the analysis of time domain NMR data obtained from static and dynamic food systems, including advanced multi-way chemometric approaches such as SLICING[10].

3 PRINCIPAL COMPONENT ANALYSIS

The basis of most chemometric algorithms is Principal Component Analysis (PCA) which has already celebrated its 100-year anniversary[11], still in splendid shape. PCA can be considered as the first amendment in exploratory analysis due to its extraordinarily robust data reduction and data presentation capabilities. In PCA the multivariate (spectral) data set is resolved into orthogonal components whose linear combinations approximate the original data set in a least squares sense. Common to bi-linear models is that an entire matrix, with each row being the measurement from one sample, must be acquired. The algorithm works on this entire set instead of on one sample at a time.

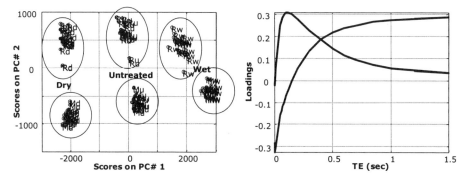

Figure 2 *Principal Component Analysis of inversion recovery data measured on oil seeds*

PCA remains the primary tool for initial investigations of large bi-linear data structures for the study of trends, grouping and outliers. Despite the fact that TD-NMR data rarely satisfies the bi-linear condition (underlying latent structures have a tendency to change with concentration), PCA has been proven very robust to this type of data[12]. Figure 2 shows an example of the application of PCA to TD-NMR Inversion Recovery data of oil seeds[13]. The reduction of the data matrix (117 samples × 23 inversion delays) is readily translated into two latent structures. The sample *scores* are grouped into 6 subclasses: dry (left), normal and wet (right) seeds. The upper parts are rapeseeds and the lower clusters are mustard seeds. It is also observed that two samples of the rapeseeds are atypically placed in the direction of the mustard seeds. The principal components or loadings indicate on the parts in the inversion recovery curves that are important. In this case the first loading (corresponding to the horizontal axis in the score plot) has the shape of a general inversion recovery curve, while the second loading gives the amount of modification. The degree of explanation for the total variation in the data for the first two principal components is above 99%. Thus, this highly compressed visualization of the *whole* data set gives an almost perfect window into the total variation in the data. More subtle details may be revealed by monitoring additional principal components.

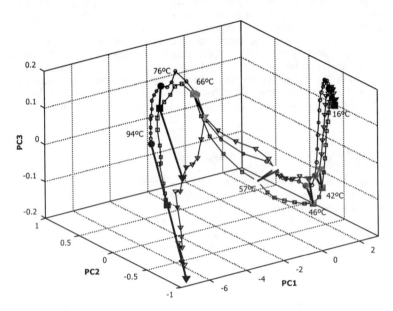

Figure 3   *Principal Component Analysis of CPMG relaxation data of three replicates of cooking meat inside the NMR magnet*

Figure 3 shows the PCA score plot of three replicates of CPMG data from a dynamic food process, the cooking of meat inside an NMR magnet. Each replicate consisting of 43 relaxation curves of 4000 data points[14]. Although the relaxation data as a function of cooking temperature is not a truly bilinear data structure (the T_2 distribution of characteristic relaxation times changes with temperature and the NMR signal decreases with temperature due to less favourable Boltzmann distribution), PCA analysis of TD-NMR data has proven to be a robust and valuable data reduction and enhancement method. In Figure 3 each point represents a full relaxation decay curve with 4000 data points. It demonstrates in full that LF-NMR measurements can serve as a window into a food matrix, as changes can be observed in a non-invasive manner. In the 3D PCA score plot the temperatures where major changes in the states of water in meat occur are readily identified; sharp bends in the scores plot, corresponding to significant changes in the properties of the water, can be identified at 46 °C and 66 °C, while softer bends are found at approximately 42 °C, 57 °C and 76 °C. These transition temperatures, identified by the PCA, can be explained by changes in the main constituents of meat during cooking: denaturation and shrinkage of the myofibrillar proteins and the connective tissue, which affects the water-holding capacity/distribution of water in meat. The lowest transition temperature found in the PCA score plot (42 °C) corresponds to the onset of myosin denaturation. The denaturation of sarcoplasmic and myofibrillar proteins begins at lower temperatures, but the rate and amount of denaturation increases significantly above this temperature. The transverse shrinkage of myofibrils starts at 45 °C, which is where one of the major changes is seen in the PCA score plot. The transition of the states of water at 57 °C can be explained by the fact that shrinkage parallel to the myofibrils starts at this temperature, but the onset of collagen denaturation beginning somewhere between 50 and 60°C could also contribute to this transition. The longitudinal contraction intensifies at 64-65 °C, which is where a major change is found in the NMR data, as shown by the PCA analysis. The intensified contraction is related to a major peak in thermal denaturation of the endomysial, perimysial and epimysial collagen around these temperatures. The slight transition observed around 76 °C may reflect further gelatinisation of collagen, which occurs up to approximately 80 °C, but it is more likely due to the onset of actin denaturation, which has been reported to happen around 80 °C. This straightforward interpretation of the transitions in the PCA results on the basis of denaturation of major components of meat goes a long way to confirm that the information contained in TD-NMR data obtained during cooking of meat is indeed related to important changes of the meat matrix due to heating[14].

If it is desirable to further compress the data into only two dimensions - the optimum for human perception - PCA can be further reduced using a so-called topology-preserving non-linear Sammon projection[15]:

$$\min \quad E = \frac{1}{\sum\limits_{i<j}^{N} D_{ij}} \sum\limits_{i<j}^{N} \frac{\left(d_{ij} - D_{ij}\right)^2}{D_{ij}}$$

where the distances between objects d_{ij} in new space (e.g. 2D) are optimized to mimic as closely as possible the distances D_{ij} in high-dimensional space (fixed). This type of technique is exactly what is used by geographers when they project parts of the globe onto a flat map. If we do the exercise in the meat NMR-cooking example, the result appears as displayed in Figure 4.

We now have the entire variation in the cooking data mapped onto a 2D surface largely maintaining the information about the transition temperatures: 42 °C, 46 °C, 57 °C, 66 °C and 76 °C already identified by the PCA. However, the directional information is lost, for which reason the Sammon map scores cannot be used in for example regression problems.

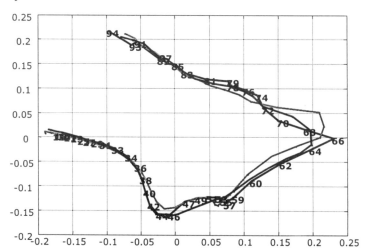

Figure 4 *Sammon map of the NMR cooking*

We have also studied a more exotic use of PCA namely Principal Phase Correction (PPC)[10]. TD-NMR signals are normally recorded as quadrature data that are magnitude-transformed to correct for phase errors in the acquisition. This procedure represents a simple and robust transformation independent of detection of the phase angle. However, it introduces a bias in terms of non-exponentiality in the magnitude-corrected data. This poses a serious problem to all data analytical algorithms based on underlying exponential structures. To solve this problem we have proposed a new, simple and non-iterative procedure for performing phase correction for TD-NMR CPMG data called Principal Phase Correction (PPC). It is based on rank reduced singular value decomposition (SVD), a technique equivalent to PCA. If **a** and **b** are two vectors of length J (Figure 5) holding the two quadrature channels, then PPC-rotation is performed by a singular value decomposition (SVD) on the matrix holding **a** and **b** in its columns. Ideally, the second score value is zero, if there are no changes in phase throughout the measurements. In practice, though, minor differences are observed and the second principal component represents noise. The product of the loading times the score of the first principal component provides an optimal representation of the phase-rotated measurements and the influence of this noise is reduced. An example of PPC-correction of a CPMG relaxation curve is shown in Figure 5. PPC extracts the optimal single-channel representation of the data, obtained in a least squares sense, taking into account that errors appear in both original axes.

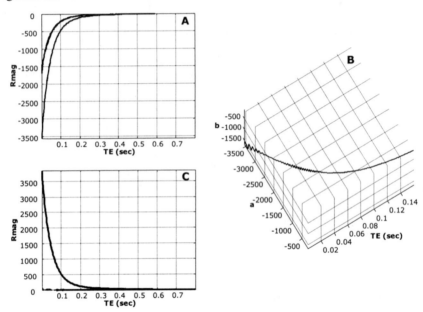

Figure 5   *Principal Phase Correction. The two channels **a** and **b** (A), **a** versus **b** (shown in the time interval 0-0.16 sec) (B) and the PPC corrected relaxation curve (C)*

4   PARTIAL LEAST SQUARES REGRESSION

Partial Least Squares Regression (PLSR), based on the PCA concept is its counterpart for regression analysis. While PCA can be compared to shopping (in the data) without a shopping list (i.e. the data analysis is performed without the use of *a priori* knowledge), PLSR is like shopping *with* a specific shopping list. PLSR is a predictive two-block regression method based on estimated latent variables and applies to the simultaneous analysis of two data sets on the same objects (e.g. spectra and physical/chemical tests). The purpose of PLSR is to build a linear regression model that enables prediction of a desired characteristic from a measured multivariate signal. It has been and still is common practice in TD-NMR to design calibration models from univariate signals collected from intrinsically multivariate signals. The most prominent examples are solid fat determination based on the ratio between the signal at 11 μs and the signal at 70 μs and fat determinations based, for example, on the amplitude of a single gradient simulated echo. Such calibrations are, from a data technological point of view, the worst approach and it is most surprising that it is still common practice in the new millennium. The main argument for not using such univariate calibration approaches is that there are no means to automatically detect outlying measurements caused by instrument drift and erroneous handling of the sample. It is, however, surprising that the univariate practice has persisted for so long and indicates that either TD-NMR instruments are extraordinarily stable over time, that re-calibrations are performed frequently or that the TD-NMR instruments can easily achieve the required precision without making use of all the measured information.

Fats are important food constituents and are measured in a number of foodstuffs to assure product quality. TD-NMR is sensitive to proton mobility, which is why the technique is useful for solid-fat determinations in the edible oils industry and for swift and accurate measurement of total fat content in dry foods. In food the liquid phase is usually dominated by two proton-rich components: water (H_2O) and fat (*poly*-CH_2), which in practice has proven to be difficult to differentiate by standard pulsed NMR experiments, as the relaxation rates of the two types of protons only differ by a factor of two in many products. The established standard NMR approach to measuring total fat in dried food (including meat) is a univariate approach where only one point is acquired from each sample, namely the top-point of a spin-echo[16]. Simple linear regression is used for calibration. This procedure requires that the samples must be dried prior to analysis in order to remove the interfering signal contribution from water protons. Furthermore, the samples are heated to ensure that the entire fat phase is liquid upon NMR analysis to obtain the total fat signal. The second evolution in total fat determination of foods by TD-NMR occurred when pulsed field gradients[17-19] were introduced to suppress the water proton signal, leaving mainly the fat signal. In principle, the pulsed field-gradient experiment consists of two gradient pulses of which the first "labels" the protons according to their local magnetic field and the second creates a gradient-stimulated echo. In this manner the NMR relaxation is sensitized to proton diffusion, as all protons that have diffused to a new magnetic environment will not be refocused.

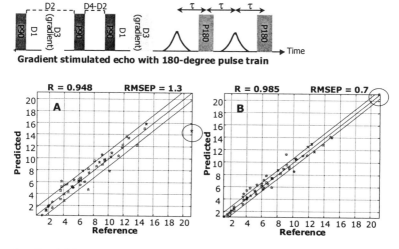

Figure 6   *Partial least squares regression. Predicted versus measured plots based on Linear regression of the gradient-stimulated echo (A) and the 1PC PLSR model of the gradient stimulated echo followed by a 180 degree pulse train (B)*

Since water protons diffuse more readily than fat protons, the water proton signal can be suppressed gradually (beginning with the most mobile) by a pulsed gradient of appropriate length and strength. By using field gradients to discriminate between the water and fat protons, the samples need not be dried prior to analysis, reducing the time of analysis significantly. However, the samples must still be heated in order to ensure that all fat components in the sample are liquid. In this method again only one point is acquired, as in the spin-echo experiment, namely the top-point of the gradient stimulated echo.

Figure 6 demonstrates clearly how a simple multivariate upgrade of the gradient-stimulated echo experiment can enhance calibration performance[20]. To the left is a calibration of the gradient-stimulated echo amplitude of 51 fresh meat samples regressed on conventional SBR (Schmid, Bondzynski and Ratzlaff) solvent extraction method. The cross-validated calibration, results in a correlation of $r=0.948$ and a root mean square error of cross-validation (RMSECV – average prediction error) of 1.3 %(w/w) fat, and it clearly demonstrates that the linear relationship between the echo amplitude and the SBR fat determination is broken above 20%(w/w) fat. To the right the corresponding calibration is shown for a multivariate upgrade of the gradient-stimulated echo pulse experiment in which the gradient echo is simply followed by a 180 degree pulse train. The cross-validated calibration based on only one PLS component has now improved the calibration performance to only one half the error of the traditional approach with $r=0.986$ and RMSECV=0.69%(w/w) fat. Moreover, this simple PLSR model is now able to linearly extrapolate measurements above 20%(w/w) fat. This type of performance enhancement is the rule rather than the exception when comparing uni- or oligo-variate approaches with multivariate approaches. For this reason it is intriguing that official standard methods for fat determination are still based on univarite reasoning – we can predict that these dinosaurs soon will be overtaken by new improved company standards, as has been the case with NIR models.

At one point in time we have also identified an exotic use of PLSR which at first appeared to be an extraordinary good idea. If we consider the partial PLSR optimization criterium:

$$\left\| y - Gb \right\|^2 = 0$$

and let the response vector **y** be a TD-NMR CPMG relaxation curve and the matrix **G** be a matrix containing a larger number of pure mono-exponential curve (see Figure 7). Then the regression coefficient vector **b** should in principle carry the T_2 distributed spectrum. While the simplicity of this method is appealing the results produced are incorrect.

Figure 7 shows a typical result of applying PLSR to the distributed exponential problem. In this case the experimental data is the CPMG curve measured on meat. The main variability in this PLSR model is the number of pure exponentials to include in the G matrix and the selection of these to obtain optimal condition number of PLSR components to include in the model. However, independent of the choices made it is not possible to obtain reasonable T_2 domain spectra. The reasons for this negative result of a model-free approach are multiple. In reality the problem is a so-called ill-posed matrix inversion problem – the matrix G which ideally should have full rank is in practise of low rank. A number of techniques have been developed to solve this type of numerical problem most of which introduces constraints such as regularisation, smoothing and non-negativity[21,22]. Obviously the condition number of the G matrix is crucial to the quality/validity of the solution and optimization of the condition number of the G matrix can be increased by, for example, logarithmic sampling of pure mono-exponentials.

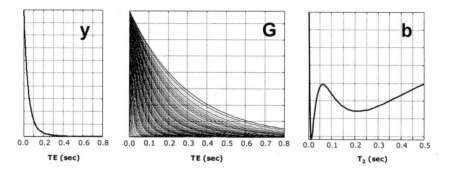

Figure 7 *Partial Least Squares Regression coefficients distribution. Non-successful 5-component PLS regression model fitting pure mono-exponentials to an experimental CPMG relaxation curve measured on fresh meat*

Despite this negative outcome, we have decided to show the result, since it includes some useful observations. Firstly, the resulting regression vector, **b**, has a peak at about 50 ms which is to be expected from a meat spectrum. In scrutinising the **b**-vector, a minor shoulder component can be observed just below 200 ms, also in good accordance with our TD-NMR relaxation understanding of meat. Secondly, although PLSR is not optimized to give interpretable regression coefficients (it focuses on the prediction of **y**), it is encouraging that the result gives a smooth T_2 distribution with reasonable trends, but clearly needs additional constraints such as non-negativity.

While distributed exponentials are often preferred to multi-exponential data analysis, because they theoretically reflect the underlying physics better, the existing algorithms are indeed most labile and the results retrieved strongly dependent on the numerical settings and choices required for the analysis. In addition, the distributed analysis is strongly dependent on proper phase correction of the data. Figure 8 displays the distribution analysis of meat cooking inside the TD-NMR magnet when analysed by standard software implementing the regularisation technique[21].

In this case the distributed analysis is relatively robust and consistent, as the sample remained in the magnet throughout the experiment and "time-slices" of T_2 distributions display a continuous development. Data has been pre-processed by phase rotation (A) and magnitude transformation (B), and the result of discrete fitting has been superimposed on the plots. The data presented turned out to be problematic to analyse using discrete multi-exponential fitting, as the system slowly changes with temperature from a two-component to a three-component system, much to the displeasure of the discrete fitting algorithm. The residuals using a two-component solution increase from around 40 °C, but a valid three-component model cannot be calculated before 59 °C has been reached (where the distribution of water is affected by the onset of collagen denaturation and lateral contraction of myofibrils, as described earlier in the paper). Thus, as the actual rank of the system is difficult to determine, the results of bi- and tri-exponential fitting overlap in the temperature interval 40-60 °C in Figure 8. Since no assumptions are made regarding the number of components in the system of distributed exponentials this approach should presumably be advantageous in the case described.

The effect of pre-processing on the result of distributed exponentials is, however, obvious from Figure 8. The results from phase-rotated data and magnitude-transformed data are very different, the problem being that they may lead to different conclusions about the changes in the meat system upon cooking. According to the distributed fitting of the phase-rotated data (A), a third water population with slower relaxation characteristics than the water in raw meat is developed around 30 °C. The relaxation time and magnitude of this population increases with temperature, which leads to the "easy" interpretation that the new population is water expelled from the myofibrils and the intermyofibrillar space to the outside of the meat matrix.

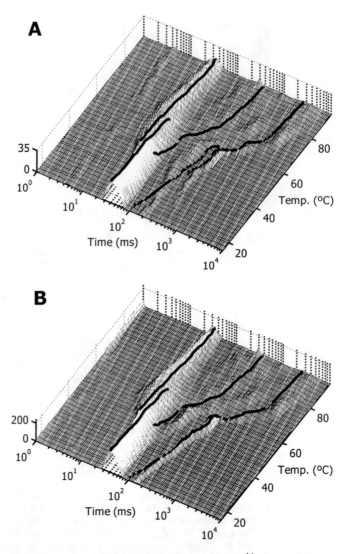

Figure 8 *Cooking meat inside the TD-NMR magnet[14]. Mesh plots of distributed exponentials calculated from the NMR transverse relaxation of meat as a function of cooking temperature*

The distribution analysis of the magnitude-transformed data, however, indicates that the third component emerges around 42 °C with an intermediate relaxation time between the two original water populations. Thus, the water in the new water population developed is relatively free compared to strongly bound water within the myofibrils (T_{21} in raw meat), but it is isolated (on the NMR time scale) from the slowly relaxing water between the myofibrils (T_{22} in raw meat). As the new population develops when myosin denaturation starts at 42 °C it is possible that the third water population represents water emerging in small compartments developed inside the myofibrils between denatured myosin units where it is trapped in a gel-like structure. Furthermore, it can be seen from

Figure 8 that the algorithm of distributed exponentials breaks down above 59 °C when the proton relaxation is incomplete (the signal does not reach 0), whereas the discrete fitting using three components can be calculated.

Most surprisingly, the discrete fitting of the same data appeared almost independent of the pre-processing and displayed the best agreement with the distributed analysis of the magnitude transformed data. This fact has led us to a dual hypothesis concerning the new water component (compartment) arising during the meat-cooking experiment. Either the new water component with an intermediate average T_2 time arises due to the formation of a porous myosin gel or is it simply just expulsion from the myofibrillar lattice[14].

5   TRILINEAR CHEMOMETRIC MODELS

Where bi-linear chemometric models provide a tremendous advantage to univariate models, tri-linear data offers an additional unique resolution of underlying components in the mathematical sense[23]. Approximate tri-linear data follow the model:

$$x_{ijk} = \sum_{n=1}^{N} a_{in} b_{jn} c_{kn} + e_{ijk}, \quad i = 1,..,I; j = 1,..,J; k = 1,..,K$$

Tri-linear models extract common loadings for all samples, as is the case with bi-linear models, but they require a cube of data, in contrast to the table or matrix of data in the bi-linear case. Several different algorithms are available for tri-linear modelling including *Generalised Rank Annihilation Method* (GRAM)[24], *Direct Trilinear Decomposition* (DTLD)[25], which is a generalisation of GRAM, and *Parallel Factor Analysis* (PARAFAC)[26] all essentially fitting the model above. While the eigen-based algorithms GRAM and DTLD providing fast, analytical solutions, they do not give the least squares solution. However, for large data sets with high signal-to-noise ratio and little model error, the difference is often insignificant. PARAFAC[27] is an alternative iterative procedure that tends to be relatively slow, but has certain advantages, including a least squares solution to the problem under investigation. Providing that approximately tri-linear data can be measured, tri-linear models provide so-called computer chromatography which is a highly attractive alternative to actually performing physical chromatography. This has for instance been shown for fluorescence excitation-emission landscapes[28].

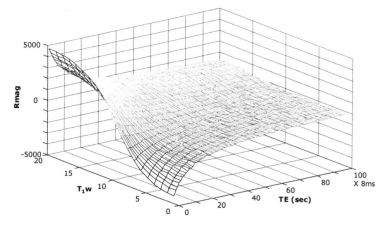

Figure 9   *2D TD-NMR. T_1-weigthed CPMG landscape of a butter sample*

Similar to fluorescence spectroscopy, it is possible to record TD-NMR data as a function of two or more variables. One such example is T_1-weighted CPMG data which have the dimension (samples × T_1 delay × T_2 relaxation). Figure 9 shows an example of a butter sample measured with a T_1 weighted CPMG pulse sequence[29].

Figure 10 shows the result of the PARAFAC decomposition of 32 samples of different spreads. The solution shows that the 2D TD-NMR data are approximately tri-linear, as loading **1** contains the *pure* CPMG profiles of three components and loading **2** contains the *pure* inversion recovery profiles. However, the third component of loading **2** has the shape of a compensation profile. In loading **3** (called the sample *score*) the concentrations of the three components of the spreads are obtained. These scores can be directly interpreted, as they require no further regression, but just scaling to represent concentrations. The scatter plot of the score of component **1** to a reference determination of total water content gave a direct correlation of r=-0.94. This preliminary investigation shows that it is indeed possible to generate approximate tri-linear data by TD-NMR and the results are most promising for future applications.

Another approach for generation of tri-linear data with TD-NMR is the so-called SLICING technique which has recently been described in detail by Pedersen *et al.*[10]. The novel approach is to upgrade a one-dimensional relaxation curve to become a pseudo two-dimensional structure and thus facilitate the unique advantages offered by trilinear models. The method is based on the fact that two different time "slices" of a given multi-exponential decay curve consist of the same underlying features (*quality:* characteristic decay times), but in a new linearly related combination of amounts (*quantity:* concentrations or magnitudes).

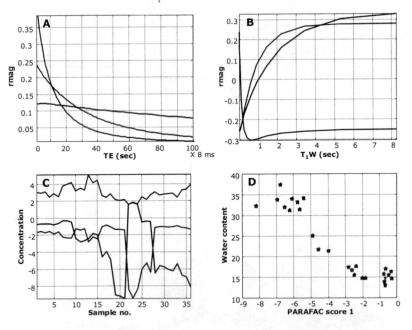

Figure 10 *Tri-linear TD-NMR data. A three-component PARAFAC solution to T_1-weighted CPMG data of 32 butters and spreads: A) loading 1 corresponding to the CPMG profiles , B) loading 2 corresponding to the T_1 profiles, C) loading 3 or scores corresponding to the concentrations, D) a simple scatter plot between water content of the samples and the score of component 1*

Windig and Antalek[30] originally conceived the idea and proposed a fast alternative to the tri-linear least squares solution, which they called Direct Exponential Curve Resolution Algorithm (DECRA). In other words, the new approach is based on the linear relationship between exponentials:

$$\exp(\frac{-t}{T_{2n}}) \propto \exp(\frac{-t+\Delta t}{T_{2n}})$$

By a simple pre-operation, multi-exponential transverse relaxation curves can be pseudo upgraded to become tri-linear data structures. In the simplest case one relaxation curve can be translated one data point, called *lag* 1, and added in a new direction called *slab* (*slab* 2), creating a data matrix with the dimension two in the *slab* direction and the dimension N-1 in the lag direction. Hence, the major part of the relaxation curves will be nearly identical, but shifted "horizontally" by a fixed amount. The idea of "cutting" data into a number of overlapping slices has given rise to the name selected for this approach: SLICING. If this operation is performed on a series of samples, it is possible to obtain a tri-linear structure that can be analysed by, for example, PARAFAC. The result of this procedure will be exactly the same as that of a discrete exponential fitting algorithm, fitting common characteristic time constants to a series of samples[10]. Extensive simulations of TD-NMR data revealed that the SLICING approach was comparable, but not superior to a robust classical numerical approach. However the SLICING approach is very appealing, because this algorithm utilising highly redundant information and requiring no initial value guesses provides non-iterative and unique solutions with perfect mono-exponential loadings. In practice, the dramatically improved speed (independent of number of components extracted) and somewhat improved diagnostics (unique solutions) of this algorithm should be used as pre-processing (super-qualified initial guesses) to traditional numerical curve resolution algorithms.

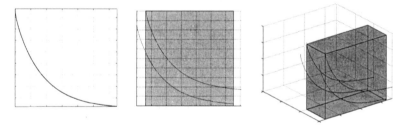

Figure 11 *DOUBLESLICING. The pseudo-upgrade of a single TD-NMR relaxation curve to become a tri-linear data structure*

It is possible to perform the SLICING pre-transformation twice (or more) and thus from a single relaxation curve generate a tri-linear data cube. We have recently conducted a series of preliminary experiments using this approach. The data pre-processing strategy called DOUBLESLICING provides a means to take advantage of tri-linear models when decomposing a single TD-NMR relaxation decay. Figure 11 shows in schematic form the simple principle behind DOUBLESLICING. In the first SLICING operation a number of *pseudo*-samples are created which all have the same underlying relaxation components, but in different relative amounts (see proportionality equation in the previous section). The number of *slabs* (*pseudo*-samples) must be equal to or larger than the number of components one wishes to extract. This *pseudo*-sample matrix is then sliced again using the SLICING pre-transformation to obtain a *pseudo* tri-linear upgrade of the *pseudo*-sample matrix. When this DOUBLESLICING is performed, the tri-linear data structure can be analysed using multi-way techniques such as PARAFAC.

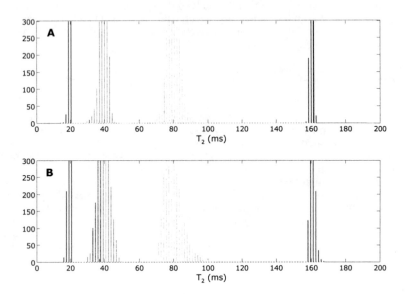

Figure 12 *Benchmark of the* DOUBLESLICING *approach on theoretical data. Model generation on 3000 replicates containing different (randomized) amounts of four T_2 components at 20, 40, 80 and 160 ms and added 0.1 % random noise. The dispersion of the solutions from the A) discrete exponential fit and from B)* DOUBLESLICING

Admitted, it may appear to be an unnessesary complication to create a 3-way data matrix with highly redundant information from a single relaxation curve to solve the simple discrete exponential fitting problem. However, with suitable resampling DOUBLESLICING produces practically the same results as traditional exponential fitting, but with significantly shorter computation times, especially when more than two components are extracted. Figure 12 shows an example of repeated deconvolution of a simulated data curve based on four exponential decays with random noise added. In this case DOUBLESLICING performs similarly to the exponential fitting, but with a computation time of less than one-fifth of the computation time of the exponential fitting. We iterate that the perspectives of this technique are primarily to be seen in cases where many oligo-exponential equations are to be solved or to serve as pre-processing (super-qualified and relatively un-biased initial guesses) to traditional numerical curve resolution algorithms.

6 CONCLUSIONS

The potential advantages of implementing spectroscopic sensors for quality control directly in the food process will create a continuous quest for still more informative and multivariate sensors to be developed. High-resolution nuclear magnetic resonance is probably the most successful and versatile spectroscopic technique yet developed and although its implementation as an on-line monitoring tool is severely hampered by the requirement of a strong homogeneous magnetic field, we foresee that this technique will also invade the more advanced segments of the food and medical industries for quality control.

The success of implementing NMR to process control is, however, dependent on qualified data analysis. With the use of chemometric techniques such as PCA, PLS and PARAFAC the multivariate nature of the NMR data can be fully utilized, increasing the stability of calibrations, making outlier detection possible and allowing for the measurement of multiple quality parameters simultaneously. We have demonstrated that TD-NMR has the capability to provide complex multivariate and multi-way information on food samples that allows application of tri-linear data analytical methods to recover pure analyte concentrations and to explore the co-variances with food quality.

In this paper we have illustrated the advantage of using multivariate data in the exploratory analysis of TD-NMR measurements on oil seeds and meat being cooked inside the instrument. We have presented an exotic use of PCA namely the Principal Phase Correction of quadrature data. We have demonstrated the advantage of using **multiway** multivariate TD-NMR data, for example by a pseudo upgrade of bi-linear transverse relaxation data to become tri-linear in an algorithm called SLICING. This method is much faster than traditional methods in extracting characteristic relaxation times and the method has for example great potential for **mathematical contrasting** of MRI images.

Perhaps most promising we have identified that T_1-weighted CPMG data can have approximate tri-linear structure and thus can be mathematically separated into its pure T_1 and T_2 components and the related concentration that require no further regression except for scaling to **one** know standard. If successful, further developments of this **mathematical chromatography** approach have the potential to separate components that are very close in characteristic relaxation times.

In this paper we have chosen to focus on the versatility of applying chemometrics to TD-NMR data. For this reason we have not mentioned the important issue of method validation here. However, the importance of validation - i.e. replicate measurements, re-sampling schemes and robust prediction techniques cannot be overestimated and is together with other important considerations such as the statistical properties, numerical stability and ease of interpretation, a continuous topic in our investigations.

Acknowledgements

This research was sponsored by the Danish Veterinary and Agricultural Research Council (SJVF), Advanced Quality Monitoring (AQM), and the Danish Centre for Advanced Food Studies (LMC). Gilda Kischinovsky is acknowledged for proofreading the manuscript.

References

1. C.F.Hazlewood, B.L.Chang, B.L.Nichols, and N.F.Chemberlain. *Nature*, 1969, **222,** 747
2. M.A.Borisova and E.F.Oreshkin. *Meat Sci.*, 1992, **31,** 257
3. H.C.Bertram, A.H.Karlsson, M.Rasmussen, O.D.Petersen, S.Dønstrup, and H.J.Andersen. *J. Agric. Food Chem.*, 2001, **49,** 3092
4. D.Johnels, U.Edlund, H.Grahn, S.Hellberg, M.Sjostrom, S.Wold, S.Clementi, and W.J.Dunn. *J. Chem. Soc. Perkin Trans. 2*, 1983, **863**
5. H.Grahn, F.Delaglio, M.A.Delsuc, and G.C.Levy. *J. Magn. Reson.*, 1988, **77,** 294
6. R.E.Hoffman and G.C.Levy. *Progr. NMR Spectroscopy*, 1991, **23,** 211
7. L.G.Thygesen. *Holzforschung*, 1996, **50,** 434

8. A.Gerbanowski, D.N.Rutledge, M.H.Feinberg, and C.J.Ducauze. *Sci. Aliment.*, 1997, **17,** 309

9. S.M.Jepsen, H.T.Pedersen, and S.B.Engelsen. *J. Sci. Food Agric.*, 1999, **79,** 1793

10. H.T.Pedersen, R.Bro, and S.B.Engelsen. *J. Magn. Reson.*, 2002, **157,** 141

11. K.Pearson. *Phil. Mag.*, 1901, **2,** 559

12. I.E.Bechmann, H.T.Pedersen, L.Nørgaard, and S.B.Engelsen. Comparative Chemometric Analysis of Transverse Low-field ^{1}H NMR Relaxation Data. *In* Advances in Magnetic Resonance in Food Science. *Edited by* P.S.Belton, B.P.Hills, and G.A.Webb. Cambridge, UK. 1999. p. 217.

13. H.T.Pedersen, L.Munck, and S.B.Engelsen. *J. Am. Oil Chem. Soc.*, 2000, **77,** 1069

14. E.Micklander, B.Peshlov, P.P.Purslow, and S.B.Engelsen. *Trends Food Sci. Technol.*, 2002, in press.

15. J.W.Sammon. *IEEE Transactions on Computers*, 1969, **18,** 401

16. E.L.Hahn. *Phys. Rev.*, 1950, **80,** 580

17. E.O.Stejskal and J.E.Tanner. *The Journal of Chemical Physics*, 1965, **42,** 288

18. J.E.Tanner and E.O.Stejskal. *The Journal of Chemical Physics*, 1968, **49,** 1768

19. C.Beauvallet and J.P.Renou. *Trends Food Sci. Technol.*, 1992, **3,** 241

20. H.T.Pedersen, H.Berg, F.Lundby, and S.B.Engelsen. *Innov. Food Science Emerg. Tech.*, 2001, **2,** 87

21. J.P.Butler, J.A.Reeds, and S.V.Dawson. *SIAM Journal of Numerical Analysis*, 1981, **18,** 381

22. S.W.Provencher. *Comp. Phys. Comm.*, 1982, **27,** 229

23. N.D.Sidiropoulos and R.Bro. *J. Chemom.*, 2000, **14,** 229

24. E.Sánchez and B.R.Kowalski. *Anal. Chem.*, 1986, **58,** 499

25. E.Sánchez and B.R.Kowalski. *J. Chemom.*, 1990, **4,** 29

26. R.A.Harshman. *UCLA working papers in phonetics*, 1970, **16,** 1

27. R.Bro. *Chemom. Intell. Lab. Syst.*, 1997, **38,** 149

28. R.Bro. *Chemom. Intell. Lab. Syst.*, 1999, **46,** 133

29. D.N.Rutledge and A.S.Barros. *Analyst*, 1998, **123,** 551

30. W.Windig and B.Antalek. *Chemom. Intell. Lab. Syst.*, 1997, 37, 241

SOLID-PHASE EXTRACTION AND ^{1}H-NMR ANALYSIS OF BRAZILIAN CABERNET SAUVIGNON WINES – A CHEMICAL COMPOSITION CORRELATION STUDY

R.P. Maraschin[1], C. Ianssen[1], J.L. Arsego[1], L.S. Capel[1], P.F. Dias[1], A.M.A. Cimadon[2], C. Zanus[2], M.S.B. Caro[3], and M. Maraschin[*1]

[1]Plant Morphogenesis and Biochemistry Laboratory, University Federal of Santa Catarina/CCA, PO Box 476, 88.049-900, Florianopolis, SC – Brazil. E-mail: m2@cca.ufsc.br
[2] Casa Vinicola De Lantier, Garibaldi, RS - Brazil
[3]Analysis Center/Chemistry Department, University Federal of Santa Catarina, Florianopolis, SC – Brazil.

1 INTRODUCTION

Flavonoids and phenolic acids are plant secondary metabolites ubiquitously found in higher plants. The interest in evaluating the phenolic constituents of common foods and beverages of plant origin has continuously increased, since they content a variety of polyphenolic compounds in amounts ranging from traces to several grams per kilogram.[1, 2] Those molecules are the active ingredients in dietary plants and have shown multiple biological effects of interested related to their antioxidant properties. For example, phenolic compounds are able to scavenge free radicals on plasma, reducing the progress of some typical diseases of the modern society as coronary heart disease. [3, 4, 5, 6]

Over the last years several studies have focused on quali/quantitative analysis of flavonoids and phenolic acids of grapes which are partially extracted from skin, pulp and seeds during winemaking. The typical (poly)phenolic fingerprint of any red wine, for instance, is determined by several (a)biotic factors in both pre- and post-harvest such as soil composition, climate, variety, viticulture practices, vinification techniques, storage and aging. [2, 5, 7] Thus, red wines are very complex matrices in respect to their phenolic composition and this feature may changes over their shelf life. [8, 9]

In order to better understand the complexity of wine phenolic composition several analytical methods have been developed which commonly used liquid chromatographic (LC) techniques, mostly high-performance chromatography, and gas chromatography (GC), coupled to UV-visible, photodiode array, fluorescence, electrochemical, or mass spectrometry (MS) detector system. These methods have allowed the analysis of several constituents of grape extracts and wines [2, 5, 7] but in general are very time consuming and expensive. On the other hand, more recently and due to meaningful technical advances, i.e. instrumentation and improvement in computerized handling of data, nuclear magnetic resonance (NMR) spectrometry has become the single most powerful form of spectroscopy in chemical and biological investigations. [10, 11] In fact, modern NMR techniques have provide a set of suitable tools to study the chemical composition of complex matrices, i.e. red wines, also in combination with LC and MS systems. [10, 11, 12, 13] In addition, further chemometric analysis of the ^{1}H-NMR spectra dataset using multivariate statistical methods has been successfully used in the classification of red wines and for the identification of beers according their geographic origins, for instance, expanding the possibilities of use of NMR in food and beverages chemical investigations. [13, 14]

In Brazil, Cabernet Sauvignon (CS) red wines are mostly produced in Southern region (Serra Gaucha) where climatic conditions (precipitation and insolation, e.g.) are expected to vary over harvests in any extent, especially during growth and ripening fruit stages. This might leads to meaningful changes in chemical constituents of that beverage, with obvious negative effects on its quality. Further, climate changes hamper the settling of a standard of chemical composition *(fingerprint)* for Brazilian CS red wines to be used as reference in further comparative studies, aiming at adulteration and quality control analysis. In order to gain more insights as to qualitative changes in phenolic composition of CS red wine, authentic samples were collected over three vintages (1986, 1987 and 1995) and investigated by using solid-phase extraction (SPE) and ^{1}H-nuclear magnetic resonance (^{1}H-NMR), followed by chemometric analysis of the ^{1}H-NMR spectra dataset.

2 METHOD AND RESULTS

2.1 Solid-Phase Extraction of Cabernet Sauvignon Red Wine Samples

Three authentic samples of CS red wine were kindly furnished by *Casa Vinicola De Lantier – Divisao Bacardi-Martini do Brasil*. The samples were originated from Serra Gaucha, a traditional winemaking region in South Brazil, and were kept protected against sunlight. After opened the samples were stored into refrigerator ($4°C$) and the internal atmosphere of the bottle (headspace) was replaced by N_2 gas.

Since CS red wine matrices are complex and polyphenolics occur at low concentration, is usually required a pre-treatment of the samples for the extraction and concentration of those compounds previously to analysis. For that, the phenolic constituents were extracted from the CS red wine samples by using a reversed-phase (RP) ODS-C_{18} classic cartridge for manual operation (BakerbondSpe™, J.T. Baker, Deventer, The Netherlands), according to reported by [15] with minor modifications. Previously to use, the cartridge was conditioned with 5 ml of methanol (MeOH) that was afterward evaporated under a N_2 stream. The CS wine sample (5 ml) was gently loaded into the cartridge and eluted using a syringe attached to the outlet with a flow rate (*ca.* 5 ml. min⁻). The extract obtained was re-loaded into the cartridge in order to elute it once more following the same procedure. A gentle N_2 stream was used to evaporate any residual solvent and the polyphenolic compounds were then extracted by eluting the (RP) ODS-C_{18} cartridge with 5 ml of ethyl acetate (EtOAc) or MeOH. The fractions were collected, the organosolvents evaporated under a N_2 stream and kept in sealed flasks at $-20°C$ for further ^{1}H-NMR analysis. All the experiments were carried out in subdue light to reduce the photo-oxidation of compounds of interest during sample handling.

For purpose of method development and data analysis, it is worth mentioning that all further ^{1}H-NMR experiments were performed using EtOAc solid-phase-extracted fractions and this is because as compounds were extracted by eluting the ODS-C_{18} cartridge with MeOH elution a low recovery of the polyphenols was found, hampering the obtainment of well resolved ^{1}H-NMR spectra.

2.2 ^{1}H-Nuclear Magnetic Resonance Spectrometry and Statistical Analysis

Prior to ^{1}H-NMR analysis, the MeOH and EtOAc solid-phase-extracted fractions were resuspended in 700 µl of acetone-d_6 and centrifuged (5000-rpm/5 min). 650 µl of the supernatant was collected and transferred to a 5 mm NMR sample tube and ^{1}H-NMR spectrometry for each sample was performed using a Bruker AC 200 spectrometer operating at 200 MHz of hydrogen frequency resonance (relaxation time 1s, excitation pulse of 90° and chemical shifts relative to acetone-d_6 – 2.20 ppm). Sixteen scans were recorded for each sample, using 16K data points. [12]

The ^{1}H-NMR spectra were further analyzed by using a chemical shift correlation matrix (*PrimGraph* v.2.0) and by means of multivariate statistical method – cluster analysis (*NTSys* v. 2.02) as previously reported. [12, 16, 17, 18] CS red wine produced in 1986 was considered as positive control for purpose of comparison due to its higher organoleptic quality pointed out by taster experts (Zanus, 2001 – personal communication) and also because more favorable climatic conditions for fruit growth and ripening were observed over that harvest.

The quality of red wines is strongly dependent on their phenolic compounds such as flavonoids, anthocyanins, phenolic acids, and stilbenes, for instance. Thus, investigations of those secondary metabolites require efficient detection methods that allow a rapid characterization of complex biological mixtures containing perhaps hundreds of constituents such as red wines. The use of analytical methods instead of organoleptic ones for the quality control of wines has continuously increased due to their performance and objectivity. [16] In this context, the application of NMR spectrometry as a detector for natural compounds in red wines seems to be a suitable approach, especially as those compounds are of interest in human health, for instance, as recently shown. [12] Furthermore, the monitoring of chemical constituents in that beverage is of great importance as grapevines are cultured in regions where meaningful climatic variations are expected to occur over the harvests as usually found in Southern Brazil.

The matrix built with ^{1}H-NMR chemical shift data of each CS red wine sample was used to calculate the correlation coefficients. The data analysis revealed that ^{1}H-NMR spectral profiles (Fig. 1) of the samples were quite similar and the results showed the existence of significant correlation coefficients ($p < 0.05$), with values of 0.89 (vintages 1986 x 1987), 0.92 (vintages 1986 x 1995) and 0.96 for the vintages 1987 x 1995 (Table 1). The higher correlation found for the samples CS-1987 and CS-1995 can be explained, in any extension, due to similarity of climatic conditions observed in those harvests, i.e. more frequent rainfalls and lesser number of sunlight hours (i.e., insolation) than in 1986. Thus, the differences observed for the total precipitation (*tp*) and number of sunlight hours (*nsh*) were about 10% and 0.7%, respectively, for the 1987- and 1995-harvests in Serra Gaucha region. Contrarily, the harvests 1986-1987 differ in the values of *tp* and *nsh* in 2.16 (116%) and 1.14 (14%) orders of magnitude, as discrepancies of 1.95 (95%) and 1.13 (13%) order of magnitude were detected by comparison of the *tp* and *nsh* data for the 1986-1995 harvests.

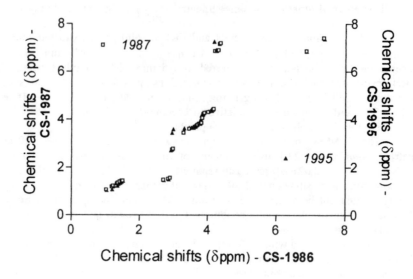

Figure 1   *Chemical shifts (δ_{ppm}) of ^{1}H-NMR spectra of the solid-phase ethyl acetate extracts from Cabernet Sauvignon red wines produced over the 1986, 1987 and 1995 vintages in Serra Gaucha, a traditional Brazilian winemaking region. The ^{1}H-NMR spectrum of the CS-1986 sample was chosen as reference for purpose of comparisons and the chemical shifts for both CS-1987 and CS-1995 samples were plotted in a correlated manner to it*

Table 1   *Correlation coefficients, association coefficients (similarity, SM and dissimilarity, DIST) and polyphenol content ($\mu g/0.1$ ml) of Cabernet Sauvignon red wines produced over the vintages 1986, 1987 and 1995 in Southern Brazil. The coefficients were calculated through ^{1}H-NMR chemical shift dataset. It is noticeable the agreement between the correlation coefficients and similarity coefficient*

Cabernet Sauvignon Vintages	Correlation Coefficient	Similarity coefficient (SM)	Dissimilarity coefficient (DIST)	Polyphenol concentration ($\mu g/0.1$ ml)
1986 → 1987	0.89	0.67	0.58	216.4 (1986)
1986 → 1995	0.92	0.72	0.54	279.8 (1987)
1987 → 1995	0.96	0.89	0.33	240.0 (1995)

In a second experimental approach, multivariate analysis using the ^{1}H-NMR chemical shift dataset was carried out by constructing a matrix to calculate the similarity (SM) and dissimilarity (DIST) coefficients, i.e. association coefficients, [18] for the CS red wine samples. The results (Table 1) proved the findings previously found by correlation analysis of the ^{1}H-NMR chemical shift dataset, as demonstrate the higher similarity coefficient (0.33). In order to confirm these results, a third series of experiment was undertaken by performing chromatographic separation of the polyphenolic compounds of the CS samples by liquid chromatography (*flash*-chromatography), followed by ^{1}H-NMR spectrometry of the EtOAc fractions as recently described. [12] Again, higher similarity coefficient was found for the CS-1987 and CS-1995 vintages, which also showed the lower dissimilarity coefficient for the measured ^{1}H-NMR spectra of the CS-samples (data not shown).

Thus, and taking into account the data obtained through either SPE-^{1}H-NMR and LC-^{1}H-NMR (*off-line*), our results demonstrate that the occurrence of homogeneous climatic conditions over the harvests had a great effect on the uniformity of the chemical composition of CS red wines produced in South Brazil. With such analytical system at their disposition, one then will be able in establish a databank over several CS red wine vintages, for instance, contributing for the characterization of that beverage produced in a given region.

2.3 Determination of Polyphenol Content

Quantitative analysis of the phenolics in the CS red wine samples was carried out using the Folin-Dennis procedure [19] and the results were expressed in micrograms/0,1 ml. A calibration curve using gallic acid (5 to 100 μg/0,1 μl), a naturally occurring polyphenol, was prepared. Interestingly, total phenolic content data found for the CS samples seem to corroborate to explain the results obtained in the SPE-^{1}H-NMR experiments, since the higher the correlation coefficient for the CS vintages, the lower the difference for the values of polyphenol content found among the samples (Table 1). Taking together, these findings indicate that phenolic composition of CS wines seems to be affected by the variations of climatic factors, mostly precipitation, observed in the production region over the vintages in study. In fact, minimum changes for the content of those secondary metabolites were observed, as climatic factors, i.e. *tp* and *nsh*, did not change significantly, mostly over the fruit growth and ripening stages.

Particularly regarding the climate *Vitis* spp. requirements, Serra Gaucha has been considered in general a humid region as the total precipitation (mm) exceeds the need of the grapevines with any frequency. The analytical strategy herein shown point to that particular climate trait of the Serra Gaucha as being of great effect on the phenolic composition of the CS wines. Furthermore, for purpose of comparison, it seems that *tp* has a stronger effect than *nsh* on the polyphenolic constituents profile of the CS wines.

3 CONCLUSION

Nowadays, the analysis of changes of chemical constituents in wines produced in Brazilian traditional winemaking regions is, among others, a basic question to be addressed in order to better characterize each sort (brand) of red wine, including both varietal (*Vitis vinifera*) and non-varietal (*Vitis labrusca*) ones. This approach will corroborate in a middle-term for the standardization process of that beverage, with positive consequences in respect to its marketing. With the aid of a multitude of experimental techniques, NMR spectrometry represents a potentially interesting technique for detailed analysis of the chemical composition of any specific wine, so that one will be able in establishing authentic characteristics for such a wine of interest, originated from any particular region (designation of origin) with typical climate. For sure, the analytical method herein described has a significant advantage as compared to organoleptic analysis for the quality control of wines due to its objectivity and performance.

References

1. J. Burns, P.T. Gardner, J. O'Neil, S. Crawford, I. Morecroft, D.B. McPhail, C. Lister, D. Matthews, M.R. MacLean, M.E.J. Lean, G.G. Duthie and A. Crozier, *J. Agric. Food Chem.*, 2000, **48**, 220.
2. M. López, F. Martínez, C. Del Valle, C. Orte and M. Miro, *J. Chromatogr. A*, 2001, **922**, 359.
3. J.E. Kinsella, E. Frankel, B. German and J. Kanner, *Food Technol.*, 1993, **47**, 85.
4. N.C. Cook and S. Samman, *J. Nutr. Biochem.*, 1996, **7**, 66.
5. J.A.B. Baptista, J.F.P. Tavares and R.C.B. Carvalho, *Food Res. Intern.*, 2001, **34**, 345.
6. R. Passos, M.S.B Caro and M. Maraschin, *Ciência Hoje*, 2001, **29**, 88.
7. E. Revilla and J.M. Ryan, *J. Chromatogr. A*, 2000, **881**, 461.
8. R. P. Maraschin, C. Ianssen, L.S. Capel, J.L. Arsego, P.F. Dias, M.A. Ricardo, A.M.A. Cimadon, C. Zanus, M.S.B Caro and M. Maraschin, *Proc. VII Brazilian Meeting on Magnetic Resonance*, 2002, **1**, 25.
9. R.S. Jackson, *Wine Science*, Academic Press, New York, 1994.
10. G.C.K. Roberts, *Drug Discovery Today*, 2000, **5**, 230.
11. J.M. Moore, *Curr. Opin. Biotechnol.*, 1999, **10**, 54.
12. M. Maraschin, R. Passos, J.M.O. Duarte da Silva, P.F. Dias, P.S. Araujo, A.C. Oltramari, J. D. Fontana and M.S.B. Caro, 'Isolation and *trans*-resveratrol analysis in Brazilian red wine by ¹H-nuclear magnetic resonance' in *Magnetic Resonance in Food Science – a view to the future*, eds., G.A. Webb, P.S. Belton, A.M. Gil and I. Delgadillo, Royal Society of Chemistry, Cambridge, 2001, pp. 136-141.
13. I.F. Duarte, I. Delgadillo, M. Spraul, E. Humpfer and A.M. Gil, 'An NMR study of the biochemistry of mango: the effects of ripening, processing and microbial growth' in *Magnetic Resonance in Food Science – a view to the future*, eds., G.A. Webb, P.S. Belton, A.M. Gil and I. Delgadillo, Royal Society of Chemistry, Cambridge, 2001, pp. 259-266.
14. R. Sacchi, 'High resolution NMR of virgin olive oil' in *Magnetic Resonance in Food Science – a view to the future*, eds., G.A. Webb, P.S. Belton, A.M. Gil and I. Delgadillo, Royal Society of Chemistry, Cambridge, 2001, pp. 213-226.
15. M.A. Delgado-Rodriguez, *Food Chemistry*, 2002, **76**, 371.
16. B. Berente, D.D.C. García, M. Reichenbächer and K. Danzer, *J. Chromatogr. A*, 2000, **871**, 95.
17. A.D. Caro, M.A. Franco, G. Sferlazzo, F. Mattivi, G. Versini, A. Monetti and T. Castia, *Vignevini*, 1994, **11**, 63.
18. R.R. Sokal and P.H.A. Sneath, *Principles of Numerical Taxonomy*, Freeman, San Francisco, 1963.
19. F. Reicher, M.R. Sierakowski and J.B.C. Corrêa. *Arq. Biol. Tecnol.*, 1981, **24**, 407.

PORTABLE, SUBMERSIBLE MR PROBE

S. Bobroff and P.J. Prado

Quantum Magnetics, Inc., 7740 Kenamar Court, San Diego, California 92121-2425, USA

1 INTRODUCTION

The food industry requires non-contaminating and non-invasive measurements to characterize processes. Nuclear magnetic resonance (NMR) and magnetic resonance imaging (MRI) have been widely used to study samples and processes in food science[1,2]. However, the applications 'in-line' and 'on-line' in the food industry are rather scarce. Many reasons account for this, mainly costs and complexity of implementation.

In this communication we describe a new, simple NMR sensor dedicated to measure relaxation times in food fluids. The principal novelty of this sensor is that it measures the NMR properties while immersed in the sample rather than having the sample in the sensor. The sensor head is compact, portable and submersible: ideal for the use in the plant where bulk fluids are to be monitored. The low resolution NMR sensor operates at 20 MHz.

2 MATERIALS AND METHODS

2.1 The magnet

The magnet piece consists of two rectangular permanent magnets of Neodymium Iron Boron blocks (high grade, 45H) yoked as depicted in Figure 1. The length of the yoke columns was calculated with standard formulas[3] to achieve the target field of 0.47 T (20 MHz proton NMR frequency). Fine shimming of the separation of the blocks (length of the columns) was performed with precision steel washers.

The magnet blocks are 5.08 cm x 5.08 cm x 1.27 cm and the yoke has been designed to allow easy access to the centre of the structure where the NMR measurements will take place.

Figure 1 *Photograph of the magnet assembly: two rectangular Neodymium Iron Boron blocks yoked together by an iron structure*

The magnet system (magnet blocks and yoke) is compact (11.5 cm x 6.5 cm x 5 cm) and light in weigh (about 1 kg). The magnetic field lines are perpendicular to the surface of the permanent magnets blocks. The magnetic field intensity has been measured (Figure 2) along the principal axes of the system, perpendicular to the magnet surface, on the centre line, with a Gauss meter.

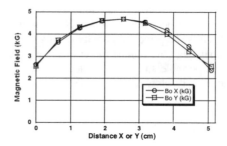

Figure 2 *Magnetic Field along the two principal axis (X and Y) of the magnet system: parallel to the surface of the permanent magnet blocks, along the centre of the structure*

At the centre of the structure, the field is homogeneous (within 1%) in a volume estimated to be 1 cm^3.

2.2 The RF coil

We have built a standard solenoid coil to transmit and receive NMR signals. The solenoid coil has 5 turns of thick, solid copper wire. The diameter is 2 cm and its length is 3 cm. The coil has been tuned to work at 20 MHz and its loaded Q is about 150.

The coil is supported by a Teflon tube (the sample fluid will be inside the tube) and the tuning/matching circuit is enclosed in a sealed aluminium box.

RF pulses are about 5 µs in length, exciting a frequency bandwidth which covers the homogeneity region mentioned in the previous paragraph. The probe requires less than 50 W to operate.

2.3 Submersible assembly

The sensor (magnet and RF coil) has been shielded with copper tape to avoid radio-frequency interference from external sources. The structure has then been confined in a Teflon box and sealed with silicone to avoid liquid contact with the parts and comply with standard food safety regulations. A Teflon tube goes through the structure allowing sample fluid to reach the region of measurement. The apparatus is depicted in the next figure:

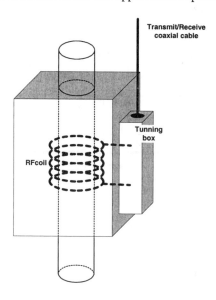

Figure 3 *Sketch of the probe in its enclosure. All parts are sealed and the tube that goes through the structure allows the sample fluid to reach the measuring region. The inner diameter of the tube is 12 mm*

2.4 Temperature Bandwidth

The system has been designed to run at room temperature (around 20 °C). The specifications of Neodymium Iron Boron magnets show that they have a strength drift of about -0.1%/°C. Thus, for the magnet assembly (2 magnets) we can expect about -0.2%/°C change in the magnetic field with temperature. Knowing that the RF pulses are about 5 μs, corresponding to a 200 kHz bandwidth, we expect that the system would operate properly with a field drift of about 100 kHz, that is within a 5 °C bandwidth. For specific applications where the temperature range exceeds that limit, the magnet can be placed in an isothermal enclosure or the probe could be retuned to work at other temperatures.

2.5 NMR measurements

The spectrometer is able to run a multitude of spectroscopy type measurements. Because of the very inhomogeneous magnetic field, we only consider measurements of T2 CPMG (Carr-Purcell-Meiboom-Gill) with very short interpulse delays (50 μs) to minimize relaxation effects due to diffusion in field inhomogeneities[4]. Due to its high filling factor, the probe is very efficient and high signal to noise is observed with standard food samples.

3 RESULTS

3.1 Static Magnetic Field Homogeneity

Figure 4 shows the Fourier transform of one echo in a CPMG experiment, which gives an estimate of the spectral line-width of the experiment. The full width at half height of the spectrum is about 70 kHz, principally due to the in-homogeneity of the magnetic field discussed in previous paragraphs.

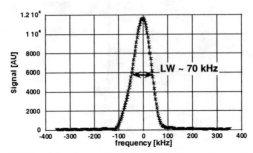

Figure 4 *Spectrum (FFT) of the first echo in a CPMG experiment for tap water. Full width at half height is about 70 kHz*

3.2 T2 CPMG measurements, T2 effective

The multi-echo CPMG measurement is fast (one train of pulses give the complete relaxation curve) and rich of information: T2 has been correlated to many physicochemical parameters in a variety of systems. The measurement of the relaxation time for the CPMG experiment in our system leads to an 'effective' T2 (T2e) which includes the diffusion in inhomogeneous fields (due to the magnet). A typical relaxation curve is plotted in Figure 5:

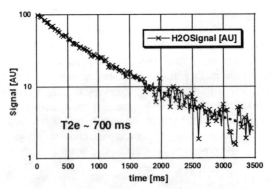

Figure 5 *CPMG relaxation curve for tap water. Pulse repetition time 50 µs, 70,000 echoes, two scans. The figure shows also the fit to the relaxation time which leads to an effective relaxation time of about 700 ms*

The relaxation decay deviates from the mono-exponential law. This is due to the effect of spins diffusing in an inhomogeneous field. We have acquired CPMG experiments for a variety of samples. Figure 6 plots a representative set of experiments. The fitted T2e relaxation times range from 900 ms to 60 ms.

These results demonstrate the ability of the sensor to characterize a large range of food fluids.

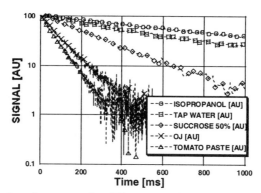

Figure 6 *CPMG relaxation curves for different samples. Pulse repetition time 50 μs, two scans*

3.3 Measurement stability

The purpose of the system is to measure slight changes in T2 for a fluid food undergoing a process. These changes in the relaxation can be correlated to parameters of the process itself. It is thus important to evaluate carefully the stability of the system. The following figure shows the T2 effective measured for a series of four hundred experiments for a 50% sucrose solution.

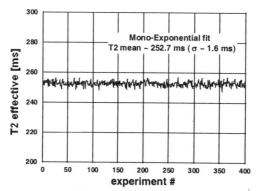

Figure 7 *T2 measurements on a 50% sucrose solution. 400 measurements, with a delay of 30 seconds between each experiment*

In this case the stability of the measurement is about 0.5% for a '2 shot' experiment (~5 seconds scan duration). Experiments on other fluids showed a similar stability as shown in

Figure 7. It has to be noted that no particular precautions were taken to thermally isolate the system, the measurements were performed at room temperature ($\sim$ 20 °C).

4 CONCLUSION

We have presented a compact, portable, submersible NMR sensor to measure proton relaxation times in fluid foods. The probe has been designed to work in the field, as a process and quality control sensor. The sensor is simple to build and inexpensive. The probe has been designed to measure proton NMR in foods, but simple modifications to the probe would allow it to work for different nuclei such as sodium, fluorine or nitrogen.

References

1   M.J. McCarthy, Magnetic Resonance Imaging in Foods, Chapman and Hall, New York, 1994.
2   B.P. Hills, Magnetic Resonance Imaging in Food Science, Willey, 1998.
3   P. Campbell, Permanent Magnet Materials and their Application, Cambridge, 1994.
4   T.C. Farrar and E.D. Becker, Pulse and Fourier Transform NMR, Academic Press, Inc., 1971.

Subject Index